ENERGY
AND THE COMPLETE
UNIVERSE.
PART 4
EU&SU
Energy and Quanta Relation.
(Thesis & Analysis) Gravity.
In Search of The Reality of Life !
2nd energy-belt
3rd energy-belt
1st energy-belt
The actual Universe.
THE SUPREME - UNIVERSE

AF530049

Title of the Book:
Energy and The Complete Universe. Part 4.
Subtitle: Energy and Quanta Relation. (Thesis & Analysis).
Gravity. In Search of The Reality of Life!

First Published:
1st January 2024

Edition:
This edition was published by the Author: Bijon Karmakar

ISBN:
978-93-341-3608-1

Dedication:
For all who seek to understand the vast Universe and our place within it! This book series delves into the realms of science, philosophy, and myth, offering comprehensive insights into the intricate fabric of existence.

Contact Information:
For inquiries, contact: [author.bijon@gmail.com]

"List of books published by the author:"

1. *The "Supreme - Universe" is the Great-grandmother of All Genesis, Part 2: The Constructional Aspects of the Baby Universe.* Published on 22nd Oct 2024.

2. *The "Supreme - Universe" is the Great-grandmother of All Genesis, Part 1: Inception of the Actual Universe.* Published on 7th Oct 2024.

3. *Energy and The Complete Universe, Part 4: Energy and Quanta Relation (Thesis & Analysis), GRAVITY: In Search of The Reality of Life!* Published on 1st January 2024.

4. *Energy and The Complete Universe, Part 3: Energy and Quanta Relation (Thesis & Analysis).* Published on 2nd Dec 2023.

5. *Energy and The Complete Universe, Part 2.* Published on 22nd May 2023.

6. *Energy and The Complete Universe, Part 1.* Published on 5th May 2023.

7. *MYTHOLOGY, PHILOSOPHY, AND SCIENCE: Book for the New Generations, Volume 2.* Published on 16th Aug 2022.

8. *MYTHOLOGY, PHILOSOPHY, AND SCIENCE: Book for the New Generations, Volume 1.* Published on 15th May 2022.

Continued from the "Energy and The Complete Universe, Part-3"

"Contents Overview"

Chapter 14

Gravity in Physics

- Gravity, Gravitational Fields, Force Fields, and Orbital Dynamics
- Energy-Field and Mass-Energy Frame Application

Gravity in Everyday Life

- Gravity's Impact on Association Football and Human Intelligence
- Natural and Non-Natural Physics Applications on Our Globe
- Gravity's Influence on Living Biological Systems

Chapter 16

Energy, Motion, and Motion-Related Items

- Motion and Its Interactions with Energy
- Energy's Role in Motion Dynamics

Chapter 17

Singularity and the Singular

- Defining Singularity and Its Application to Energy in the 2nd-Energy-Belt

Chapter 18

The Significance of 0/Zero

- Application of Zero in the Observable Universe and Energy-Universe
- The Role of Zero in the 2nd-Energy-Belt

Chapter 19

Recollection of Some Points on the Construction of the Universe

- Review and Reflection on Universe Construction Principles

Chapter 20

In Search of The Reality of Life!!

- Quantum Paradox: Being Alive or Dead

Life and Energy

- Misinterpretation of Terms like "Dead" and "Decay"
- Nirvana and Its Energy Implications

Free Will and Action

- Free Will in Bio-Intelligent Systems

Civilization and Humanity

- Technical Importance of Civilization for Humanity
- The Concept of Aliens and Extraneous Influences

Life Beyond Humans

- Survival Mechanisms and Their Impact on Physical Systems
- The Broader Meaning of Life and Its Application Beyond Humans
- Life Never Ends, and Death Never Occurs from an Energy Perspective

Humanity and Artificial Intelligence

- Interaction Between Humans and AI

Religion and Energy Levels

- Religions, Traditional Practices, and Brain System Scrambling
- Concept of "Energy-God" and Faith in God
- Characteristics of Energy-God (Based on Earlier Books Volumes I & II)

INTRODUCTION

Clarification on the equation E = mc^2 means opening the surprising reality beyond the determinable knowledge of the past that allows computation of some tomorrow and extending the ascertainable understanding for the unanticipated futures based on the precise calculation of the stored past.

Going heart and soul into this equation reveals many unknown truths that classical ways set up in many minds, allowing us to overcome and comprehend the fundamental reality of nature and the Universe.

Besides, comprehension of energy levels' functionalities can help many to overcome the probabilistic predictions that usually arise due to standard measuring practices (*i.e., mainly 3D and some reasonable supplementations*) following the daily activities of the assessable physical world.

As usual, the equation E = mc^2 provides a deeper comprehension of the observable and non-observable Universes by revealing the interconnection between energy, matter, material, microparticles, etc., whatever appears as reachable or beyond perceptible sense to intelligent species.

Its' (E=mc^2) clarification will allow for enhancing graspable capacity concerning the actual functionality at energy levels' instead of predictability based on some measurement methods restricted by the ascertained physical items. In other words, it opens up new opportunities for scientific exploration and discoveries, besides the actuality of knowing about the position and pride of intellectuality as the human species in this parentally caring habitable Globe.

The symbolically indicated equivalent relationship between Energy (E), diverse state of energy-mass (m), and speed (c), here 'c' is the motion deriving essence at different levels (^2) of quanta, and classical physics in all, which seems quite complex or simply fantasy.

At the most fundamental level, energy is quantized, meaning that it (*energy*) can only exist in discrete packets known as quanta. That is the basis of quantum mechanics, which describes the behavior of particles at the quantum level.

When energy (E) termed by specific areas like ^2 or <2> with some compositional values of mass-energy's mass (m), with multi types energy elements or working essences and speed (c) {(*or the symbol 'c' depicts as motion deriving component for a pre- or post-state mass-energy frame*)} represents somewhat an independent role since they restrain by some dimensional configuration compared to vast surroundings of energy-Ocean.

And then internally begin to engender energy physics due to the functionality of constitutionally developed energy elements following the fast-altering capacity they (*energy elements*) act, and types and quality determined based on the causal effects of internally in the mass-energy frame while encountering external atmosphere.

Fundamentally, Energy (E) maintained an eternal equivalent relationship with classical, biological, and quantum mechanisms regardless of whether intelligent beings can discern through the limitation of biologically equipped organs.

Also, at energy levels' all diversity materializes for making myriad amounts of most-micro to big objects maintaining perfect loss and gain balance in the pacific of the 2nd-energy-belt.

Chapter 14

Conscious And Consciousness.

The dictionary expression of the word conscious means something aware of and responding to the individual surroundings.

And consciousness is sentience or awareness of internal and external existence through intelligent systems.

i. Although the subject of the consciousness of the living system remains a puzzling and debatable issue for millennia, even in the modern scientific era.

ii. But, from the viewpoint of Energy, I may present in brief and say consciousness acquires naturally from the molecular level of any living unit of a species of native or alien natures whenever they attain a state-of-Stableness under the following proportionate internal and external physics values leverage.

iii. Analogously with the other living species in this bio-Globe, the intelligent human species usually recognizes them as bio-nature's avant-garde products.

Since a living being represents similar to a union of integrated millions of multitype molecules, having individuality and intellectuality due to the improved memory storage capacities like brain box and sensories that determine the awareness comfortably in decorous /recognized surroundings.

→So, from the constructional aspects, it can say that the consciousness of our intelligent systems mostly predeterminate value since all awareness-gathering efforts are in-built into a system, besides some are gifted from the ancestors through genomic bearings, and in addition, many natural and non-natural acquirements allow to improve consciousness while spending daily life in sensible surroundings.

iv. However, for the typicality of machine /mechanical way created artificial intelligence, the humanoid robot or likewise is a somewhat predetermined value, programmed or fed by natural intelligence or bio-intelligent beings to respond automatically through the constructed non-natural sensory systems. And accordingly, artificial intelligence (AI) and others are considered artificially conscious.

v. From the viewpoint of Energy, somehow-someway, everything in the Universe may regard as conscious whenever affiliated with a time-bound (*time-bound generally a measurable machine-made time, or advanced stage's energy-Time*) stable position as a state of the mass-energy frame following the mc^<2>.

{(Herewith give a little technical view about the argument part as mc^<2> and convey that whenever a unit-value mass-energy frame act as a time-bound zero-status-stable position with some reasonable amount of lower or upper level of 'm' or initiatory mass, capable of organizing and deriving its (m or mass unit's) comparable thinner degree of energy at the center to represent motion element for performing self-deriving motional-activities before it becomes a zero-status-stable mass-energy frame under a conductive physics values surrounding.)}

→Here may give an example of the common belief that a Rock or some other products outlined by classical physics, like Solids, liquids, gases, and plasma, seemingly have or genuinely do not have any consciousness.

vi. Following the contemporaneousness of the physique and satisfying frame of awareness of human beings, it may be unease to convince that everything, whatever the state of matter, classified the objects into some delimiting sections following their proprietorial values, includes a solid-state Rock, whenever attained a stable position, natural ways cultivated with some form of consciousness.

→Briefly, for clarification and following classical physics, it can say a unit-value rock is a compact noticeable physical state composed of billions /trillions of atomic and subatomic particles.

→So, the Rock responds with the applications of high or low values classical physics like temperature (*cool and hot*), pressure, coercion, etc., that impact and brings the changes at energy levels from a state of standardness in a surrounding, following the rules of classical physics.

(Observably, someone may justify that Rock cannot exert internal physics. But considering the physical aspect, like the structural and proprietorial characteristics, the Rock must retain some agreeable standard in a reasonable external atmosphere, which allows it to perform the internal energy level's physics values activities by maintaining loss and gain up to some acceptable benchmark, and so the virtually unchanging physical status of the Rock.)

vii. Now, assuming from the subatomic level of the Rock, the consciousness acquired by the microparticles while unitedly performing to preserve the status or properties of the Rock in a place of its stable position.

But upon application of non-standard external physics and gradually with the changing momentariness of the internal (*Rock*) energy-Times at the ground level, the preset overall consciousness of the Rock gets disturbed and calls for the alteration at energy level from an earlier status, following the external physics values applications.

viii. Now, for a moment, may intercept and widen the pre-set value of an individual mindset (*living being*) on classical physics and presume reasonably that the atoms /sub-atoms are independent sacks /packets of energies.

And, in a deep sense, it may be easy to understand that at the grassroots level of the unit value, 'Rock' is nothing more than the lower level or upper level of a compact state 'sack-of-energy' or 'packet of energy' with operating components like energy elements.

ix. By the way, here may give a little touch and convey that the works of in-built energy elements are the True source of physics, somewhat or some level of similarity derived as an overall functional property/s of matter defined through classical physics either of large or at the most microstate particle level's effective outcome in the form of temperature, pressure, coercion, and likewise many yet to discover through the conceivable physical worlds (*technicality of the physics derive at the elementary level called elemental-value of energy elements defined almost all earlier books*).

x. Further, concerning the above and applying technical works of electronic equipment, it may believe that electronically generated physics (*electronic physics is somewhat similar to classical physics derived from a large matter product since electronic components belong to some state of matter*), which utilizes to set up sensing organs in a humanoid robot, is the root cause of deriving artificial consciousness, mainly a predeterminate implication for designing Artificial intelligence (AI).

→However. one day, the new and advanced generations will be fully conscious of the energy level's elemental values roles, which are the True root cause of deriving all physics values regardless of whether someone of the contemporary generation can understand following the accessible physical world through the delimited design of living beings.

14.1) A little more about the consciousness loss, or casualty of consciousness to a living and non-living intelligent system, also the same applicable for rest of the energies and energy products engaged in the 2nd-energy-belt.

Generally, consciousness in a bio-living system develops during extant in endurable conducive surroundings in coordination with **information**, intercepted and transported by the constituent sensory organs to the central processing system like the brain box, then stored in memory to perform and apply for physical, sociological, and some other purposes.

*(By the way, it may give a little touch for circumventing any confusion about some technical view at the energy level, and say, the word senses '**information**' and '**experience**' have two different applications even though they are somehow interconnected.*

Information is some form of a sequence of energy signals, carrying working physics at the elementary level of energy elements.

*But the **experience** happens when something and someone encounter or undergo events or occurrences upon receiving information through some means.)*

i. To convince in briefs and easy ways on the subject, at first, I may quote some points about the 'casualty of consciousness' that happens when an individual physical system disrupts, which depends on biological fitness to the standardness of parametrically accepted normalcy in a life-sustaining nature, that is, something a value of 'constancy' for computing physical status.

ii. The disturbance of consciousness to an individual bio-being may assume a whole of monostable unit's or its part/s, loss of internal equilibrium level due to some reasoning of the weakness of constituent bio-components from an assuming healthy status while the application of external or natural physics values is in standard.

iii. Now, it may think deeper about the True position of Energy or a mass-energy frame reasonably at active, move or rest. And if someone can imagine that the physique of a human being represents an overall large mass-energy frame (mc^2). Or, all in all, a sack /packet composed of bio-grade Energy.

Then, the interruption or loss of steadiness in the standardness of surrounding/s derives due to constituent energy level obsessional roles that act on the physical position exhibiting weakness of the physique, which may lead to some form of 'casualty to consciousness' while processing regular activities in the brain box to control a bio-system.

iv. For the methodological example, it may quote that someone must have heard about the data incident in the modern computer system. Somehow this data incident may be analog to the word-sense casualty-of-data due to some malpractices that interrupt the healthy or conscious /sensible operations of a computer, machinery, artificially intelligent system, etc.

v. Following the above, the sense of 'casualty of consciousness' may apply to any mass-state energy while they are in some level of standardness regardless of rest, potential, and kinetic role.

The sense of Casualty-of-consciousness may justify involvement even in any life-serving world like this bio-world, alien world, and even to the observable, non-observable, and energy universes in the 2nd-energy-belt.

vi. Hypothetically quote, any life-supporting world is conscious about its territorial integrity to save its creations in balancing order of surroundings that each now and then tries to disturb as a place of stableness. Likewise, all observable and non-observable universes in this 2nd-energy-belt acquire some consciousness to reserve themselves at some stage of stableness in this turbulent 2nd-energy-belt.

→ {(*Due to the constitutionally set consciousness of any living, living being and other objects commonly tend to reserve stability under the standard and even some non-standard atmospheric circumstances.*

Further, in a remote sense, it somehow resolves the classical question based on a noticeable analysis of why Newton's first law stated that 'every object /body remains at rest or in uniform motion in a straight line unless compelled to change its state by the action of an external force'!

Since the physical status of a living body-mass or a living being consistently performing its dynamical roles, whether perceptibly recognized as active, feeble, or dead.

Whereas, at energy levels, the body's physics is regularly in force to work to maintain lively and stable status.

*So, the significant healthy consciousness capable of maintaining balance by applying its structural parts as limbs, organs and through internally generated physics values (*body physics*) in the standard surroundings.*)}

vii. Let me share a brief thought on the significance of 'perception' and consciousness.

By the way, it may recall one thing about the word sense perception and its augmented applications, which mainly allows for grasping the physical reality through the delimiting sensing systems, causing improvement at some levels of bodily awareness.

But it does not mean physical 'perception' will entirely fulfill energy levels' qualitative and fast degree responsiveness.

Accordingly, any small or big physical state mass-energy frame (*means, matured mass-energy frame's mono unit or complex union*) is comparably poor-quality and less conscious than the original form of energy and the immense energy -Ocean of the 2nd-energy-belt.

14.2) The technicality of consciousness.

Technically from the viewpoint of energy, consciousness is some outcome as special energy-Signals with a specific micro stage's spectrum derived deep in the elemental level of energy elements in the unit-value frame of mass-energy.

→At this moment, may consider consciousness initiates due to some unusual elemental values or physics values acts to a sensationally susceptible bio-physique, mainly for external involvement of surrounding atmospheric physics induces for many (*conceivable and inconceivable*) reasoning. That external physics consequently impacts the standard elemental physics of zero-status-stableness of the mass-energy frame, a representative of a building block or molecule.

→Some technical viewpoints concerning responsiveness to a living unit/s while in liveliness.

For an intelligent being, the True outcome of responsive energy-Signals which effectively initiates deep in the elementary level of energy elements.

This energy-Signals stepwise gets processed on the point of a system-built reality due to the diversified impact of body physics generated from integrated molecules before the final delivery through different expression techniques (*like gesture, vocal, visual, etc.*) after analyzation in the brain box.

Likewise, almost everyone and everything, those who occupy specific processing systems like the human brain box or something else as subordinated quality, more or less follow the same principle, whether evolved in natural or non-natural ways as livings or with non-living dignity.

These types of special energy-Signals with specific properties mainly serve the responsiveness to a physical system of the lively unit shaped up from specific grade/s of life-deriving energy.

→Accordingly, on the point of distinctiveness, the consciousness of bio-species adapting in this bio-Globe may seemingly alike to an alien species, but certainly not equal.

→It is easy to understand, deep from the energy level of a unit-value building block or molecular level that evolved following an evolutionary process, parametrically synchronizes in the conducive adaptable surroundings in a native nature like this bio-Globe, may find it difficult to reorganize when an attempt to adjust straightly in an unconcerned atmosphere.

Therefore, it needs a massive change of surrounding external support or the improvement of internal energy levels arrangements for a parametrically balanced complexly positioned unit-value big living.

→So, it may comprehend from the standpoint of mathematically determining differential constants, which can implement in an intelligent living system for their in-built computation system. And accordingly, it can convey that, in any suitable surrounding, a zero-status-stable molecule acts as a self-determining constant.

Also, when the molecules are unitedly working for a big lively human living unit, all their differences come out through the brain processor as an overall constant after computation /analysis while representing a zero-status-stable compact mass-energy frame.

Likewise, for responsiveness, compounded way shaped by any group of intelligent species usually analyzes overall, upon getting all differential feedback from individual zero-status-stable Constance, constancies of the group intellects, also the same principle happening whenever response yield by other livings.

14.3) Time and Consciousness.

As quoted earlier, to a general acceptance of living bio-intelligent, Time is simply a measuring tool for countable purposes as a period during which an object is in action.

→So, from countable Time, we intelligent species consciously satisfy the works of accessible items. And beyond that, many of us will deny accepting the existence of Time and its True purpose of expression.

→Think in-depth and assume there is not at all existency of Time to determine the surroundings by the constituent living biological system.

Then, the naturally evolved consciousness will find it difficult to calculate one's existence as a life form to a determinable Time-period due to the constituent biological Time, which will not accept the daily requirements for a healthy physique as peace, ruckus emotion, defense, offense, and so on undergoing roles while surviving in a surrounding through the in-built delimiting sensory faculty that will deliver on reckonable Time/s concerning constituent functionalities to a bio-intelligent processor.

→And for something like a sponge /earthworm without a brain but has consciousness for seeming not reckonable fleeting inbuilt Time only.

→Also, for non-living items without any operating processing organ like the living brain box, the acts of consciousness depend on the surrounding's naturally deriving atmospheric-Times and integrated measurable energy-Times that work proportionately for arrangement and disarrangement at the energy level to make or dismantle a zero-status-stable physically appearing object.

→So, one may say individual constituents consciously determine the Time-bound activities in coordination with the surroundings to secure life as the existence and the perceivable works in the 2nd-energy-belt. And it does not require any non-naturally determining Countable Times.

→ {**By the way, here may reemphasize the quote in the book 'Energy and the Complete Universe, part-2**.

(*Somewhat like, 'So, in a multi-mix compounded organ from any point of origin, information loss is due to Time dilation and Time-constriction, which became general practice before reaching near or far-distant processor/s set in a physique.*

It is due to the loss of momentary energy-Times that leads to information loss, beginning from a molecular level, then to compounded way-shaped organ before reaching a far distant processor for analysis, which is somewhat like an active human brain box'.)}

14.4) Some more Technical views and analytical presentation following grass root level works of energy about the consciousness on energy-Time.

Motion is the root cause of deriving reckonable 'Times' (*reckonable Times generally caused by energy-Times*) to any unit-value mass-energy frame or object and objectual product while in a state-of-Stableness whatsoever some flexible or stiff roles, capable of reserving balance **rhythmicality** on countable Time-periods in conducive surroundings.

But while judgment makes through perceptible musical sense, many of those types of energy level's **rhythmical** performances are seemingly incredible to accept since the subject is a new novelty to human consciousness.

(*Here, rhymical means regular in order by following the universal principle.*)

→In the 2nd-energy-belt many observable and non-observable items, including the multiverses, are exhausting, rebirthing on true-Times or energy-Times.

And accordingly, the essence of consciousness of any form of object keeps on transferring until the mass-energy frame dissolves and finally assembles with the ostensive Purest-energy and retains poising the eternal maturity of the 2nd-energy-belt without the excessiveness of anything, whatever the human species consciously considered standard or rare, following the appearance and disappearance of accessible items in a surrounding.

→In short, it recites that in the 2nd-energy-belt, following the differential activities throughout, ostensive Purest-energy may utilize as a True invariable constant. Or something like absolute-Zero (**abs**$_0$) for every sort of computation all over the functional energy zones holding accessible objects, energy products, and all inaccessible diversified units of mass-energy frames.

14.5) Validity of the word senses reality, locality, and likewise on the point entanglement, as well as for our consciousness.

The word sense **locality**, in common, clarifies some specific area/s to a locale of an individual or group of citizens.

And the word sense **reality** determines the items practically accessible and experienced through the bio-intelligent sensing organs directly and indirectly that reasonably accept by the background of one's consciousness.

So, from the viewpoint of energy, it is clear that expression through those word senses like **locality** and **reality** is somewhat a projected position overall granted by the constancy of bio-grade intelligent species determined through some acceptable coordinates for the confidence of measurements.

→Now, think in depth about the constitution of independent position as a bio-intelligent being, where molecules play the most significant roles from the grass root level and contribute lots to the essence of consciousness, whenever discharged after processing in the brainbox as an overall final value that subsequently tends to regulate the physical livingness of a human being.

Indeed, molecules are arranged complexly in a bio-intelligent system that directly and indirectly links throughout a body system, even if they are apparent as weird relations between the participating molecules of independently performing organs.

→Now, think in-depth about the networking ways of entanglement through the complex wiring and remote or wireless, from any assumption close to the distant molecules of a bio-physique representing a zero-status-stable positioning.

Each independent molecule solidarily occupies a specific boundary for working limits whenever acting under the naturally gifted flagship as zero-status-stableness.

So, from the viewpoint of energy components or the energy elements, each molecule directly depicts a 'sack-of-energy' or a packet of energy capable of generating at the most micro level operational signal for deriving liveliness activities of a molecule then the group way managing organs, and subsequently multiple organs represent as the whole of a living complex body system like a human being.

→Now, logically from the viewpoint of the general consciousness and considering the assessable items through the delimiting anchored sensory system of a physique, if someone says that the sense of **reality** induced from the deep in the **locale** of molecular level, that will be somewhat ridiculous to accept through the traditionally set consciousness.

In this regard, I may give an example of a humanoid robot and its performing works, which improved consciousness of bio-intelligent accept as fact and valid.

Whatsoever the invention for some purposes with tolerable variation as the positive and negative value that befit to the standardness of developed human consciousness, even though the active participation and the overall outcome of inbuilt multi-purposes use electronic components of the humanoid robot caused to derive for discernible **reality**.

→The induced **reality** of the assessable practicality initiates from the ground level in the locale of triggered molecules. Then loses the actuality of information in many ways while passing through the grids of entangled networkings before finally appearing as an overall outcome.

The below drawing EU35 assumes a molecule or organ and an acting sack-of-energy compositae with dedicated operating energy components entangled through visible networking lines and the possibility of remote links throughout a complex physique.

And the **grid points** indicated in the drawing mostly intertwine with other molecules, also are the acting trans-receiving link and the place of possible genuine information loss that derives from the origin, or the same may mean reality loss of delivered communicating information at a molecular level.

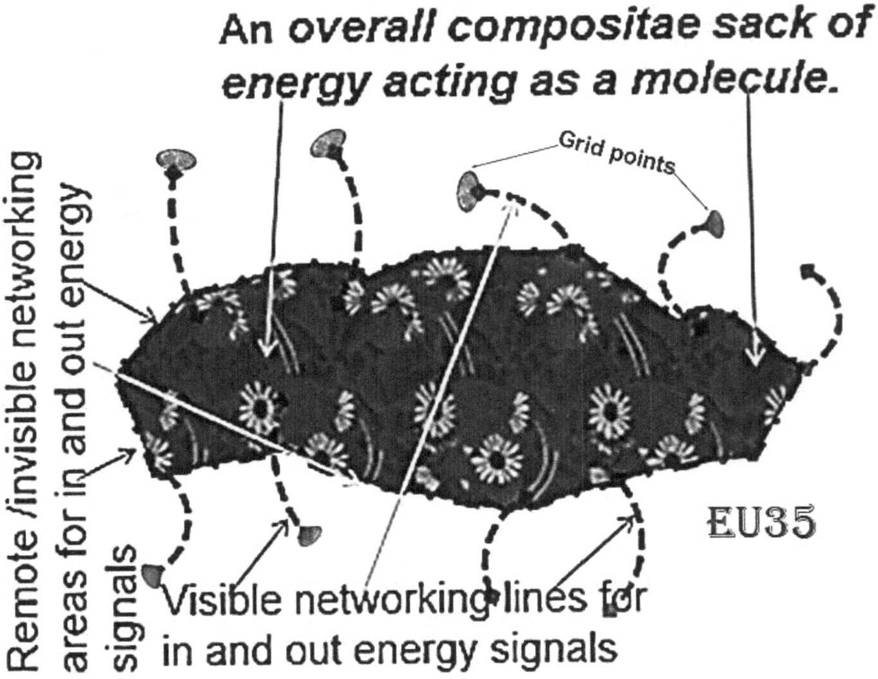

i) By the way, it may refresh by remembering some points concerning the grid points at the energy levels of energy elements and the trajectory of elemental values.

In a common conception, a grid point refers to a point of intersection where equidistant interlocking perpendicular and horizontal lines meet.

a. However, from the standpoint of energy, the observably applied sense as grid-point (*drawing EU35a and EU35b*) may not seemingly match in any event to the standard consciousness of intelligent beings since the accountable direct 3D and indirect add-on computational methods may give some satisfactory conclusion to specific ways designed living species and their creating machine works.

Despite some noticeable cases, the word sense illusion or mystery still dominates the fundamental 3D measuring strategy since three-dimensional ways measuring methods cannot surpass the fastest multi-directional physics values activities that happen at energy levels.

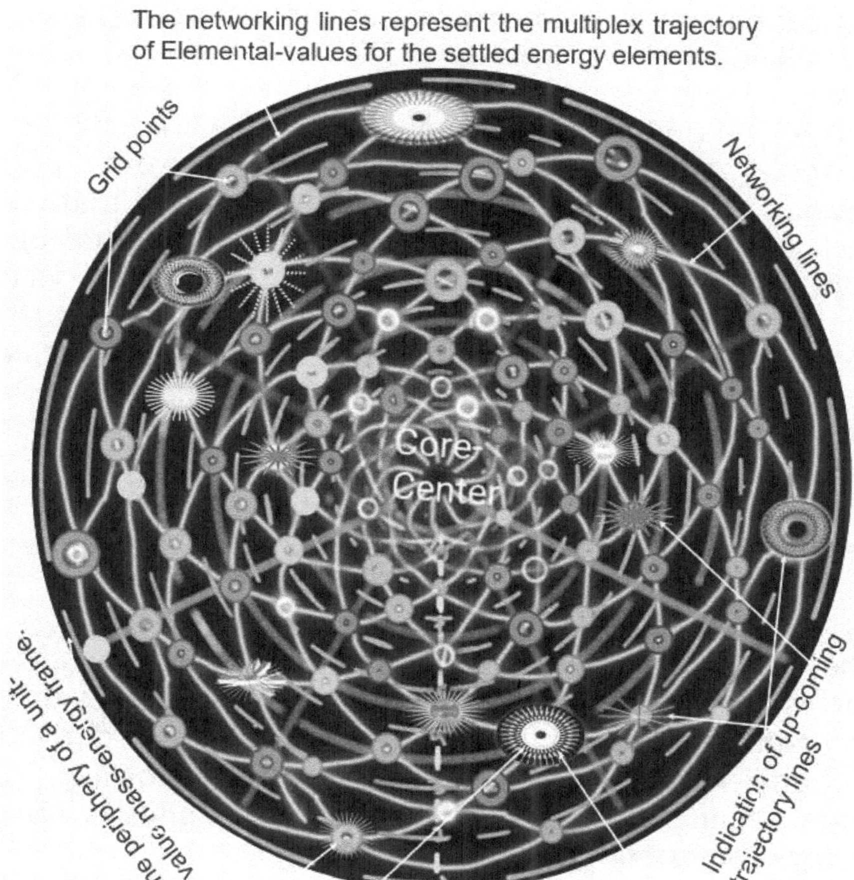

EU35a indicates a unit-value mass-energy frame display with a few short and long arrangements of intricating energy elements, set as provisional base points on individual grid points.

b. Nonetheless, the same word-sense '**grid point**' may implicate for indicating as the meeting point/s of most instantaneous diversifying energy levels work and orderly management in the small or massive mass-energy frames.

Also, multidirectional approaches are not limited to some discernible specific tracks but encompass multiple unusual directions.

c. The drawing EU35a indicates a unit-value mass-energy frame display with a few short and long arrangements of intricating energy elements, set as provisional base points on individual grid points.

Those grid points in a mass-energy framework may justify as some point of settlement or post for networking lines develop due to elemental values persistent roles that cross each other for "complexly settled energy-elements" other than the centrally situated aggregated lighter values energies (*core center*) assumed as the centrally situated commanding position of motion-element (*motion-element is unique as a mass of motion deriving component for any form of mass-energy frame*).

d. Also, in the unit-value mass energy framework, the networking lines represent the multiplex trajectory of Elemental-values for the settled energy elements.

The movements of 'elemental values' depend on corresponding variations of deviating energy values due to the consequential impact of centrally situated derivative Motional-values of motion-element following the underpinning atmosphere or atmospheric physics (*Twig+ energy atmosphere*).

e. Besides the drawings EU35b, the EU35c is a sectional view that shows almost the entire surface coverage by a grid point, indicating multi-stemming touch points and tangent paths of the elemental values of energy elements, which (*energy element*) momentarily share or settle at a comprehensible grid point in a complexly structured unit-value mass-energy.

A sectional surface view of a grid point in a complex mass-energy frame where considerable elemental values (shown as ripples and waveforms) touch /interlope and organize or reorganize by supplementing or abating momentarily in concurrence with the standard atmospheric physics.

EU35b

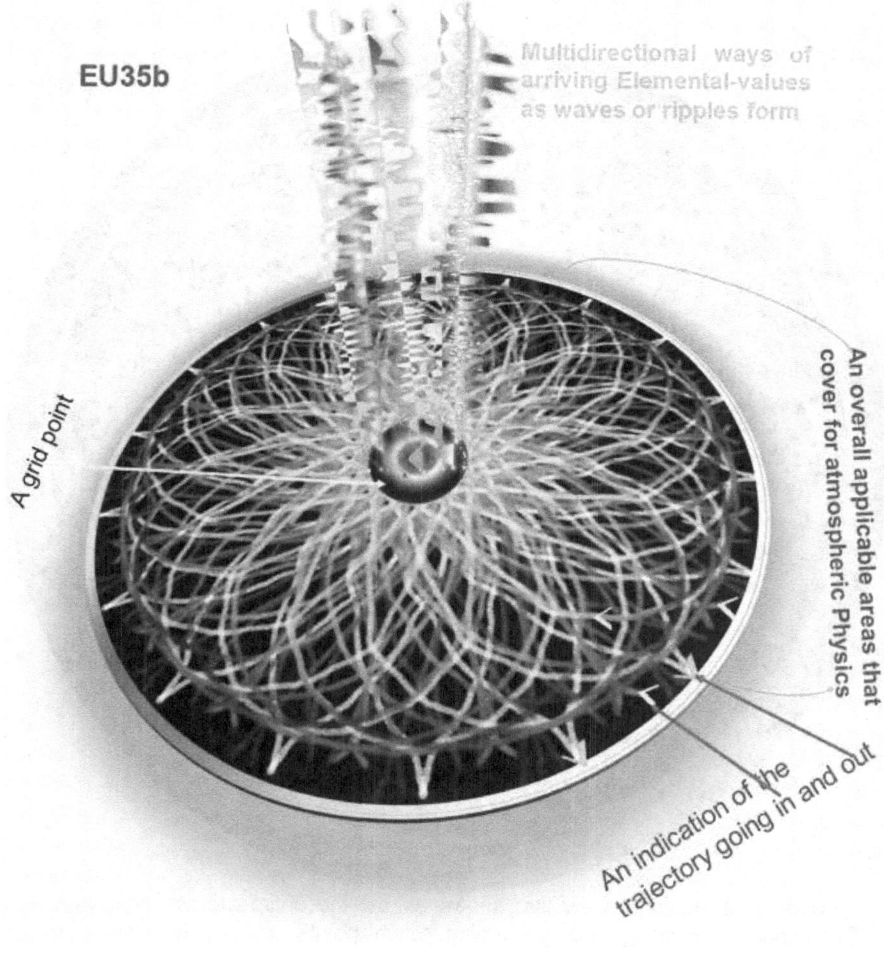

{(*Regarding the* ***energy element****, the word sense 'the most* ***momentariness****' is its true feature, whatever the energy elements perform as a pure aspect of energy element or serve (*shown in the drawing EU35a*) upon settling at a grid point resulting from multiple elemental values meeting at a specific mark.)*}

→So, in a union or mass of multiple types of energy representing an object or a big sack of energy with an overall mass-energy framework, many energy elements complexly act to reserve stability in harmony with the surrounding standard /reasonable atmospheric physics.

EU35c is a sectional view that shows almost the entire surface coverage by a grid point, indicating multi-stemming touch points and tangent paths of the elemental values of energy elements, which momentarily share or settle at a comprehensible grid point in a complexly structured unit-value mass-energy.

f. Those interiorly linked grid points also may consider base points for methodical roles of all independently positioned energy elements that settle under the excellence of the **'law of circumstances'** in the unit-value energy or their complexly compositae molded unit comprehensible as an object or zero-status-stable mass-energy frame.

(*Undeniably, for the intelligent living species and their particular ways of discernment, analogously, those fastest versatile or camouflaging roles of energy elements seemingly be fantasy to comprehend, even though it is factual for energy level functions. Since the **'law of momentary circumstances'** comes to perpetrate due to the overall influence of the natural law in the local atmosphere.)*

ii. To the consciousness of the human species, no tangible or impalpable physical appearances can justify as the fixed or perpetually settled in this 2nd-energy-belt other than the diverseness features of energy.

a. Accordingly, any object may justify as an accurate /True recognition (*means without parametrical arrangement*) while micro momentarily altering marker points from previous placement at the elementary level's roles of energy elements can perfectly measure in a unit value mass-energy frame representing the object.

So, following the prevalent versatile roles of energy elements, it is reasonable to convey that the physical reality (*means, derivement of overall or standard value due to combining efforts of internal and external elemental values or physics values outcome*) is possible when measurement or judgment can determine accurately at pin-points of elementary levels of energy elements, accumulating some amount of energy and encompasses it to make a predisposing-constancy or zero-status-stableness unit.

b. By the way, here to quote, in an assumed assessable (*that is, some physical aspects*) locale, with the extension of events on energy times or repeating circumstances or events on energy times, circumstantial law gradually transforms into broader coverage for representing natural laws that act as a benchmark to many discernible items.

(*It also means, following the fundamental 3D appraisals, the laws that justify as natural laws, mainly origin due to the endless circumstantial laws, begin with momentary works at energy levels that gradually act an overall outcome as perceptible natural laws to many living species, adapt in an inhabitable atmosphere.*)

→So, from the standpoint of energy, factually, nothing can justify as permanently settled or fixed any virtually resolved position or point following the 3D-based analytical system. Instead, it follows the conditionality of everything, which is directly or indirectly linked with internal and external activities of energies regardless of any object, mass-energy frame, and even the seemingly devoid, vacuum, or peaceful free-flowing energy in the matured 2nd-energy-belt.

14.6) Finally, from the perspective of some technical talk about the aspects of multi-dimension, and then judgment through the brain processor of human beings based on three-dimensional physical computation methods for analyzing the delivered communicating signals from the molecular levels.

Herewith a multidimensional projection (drawing EU36) following the structural importance of intelligent beings that are capable of judging the observable items, mainly in three-dimensional ways reserving the center a common axis point, body line, or center point for right, left, up, down, or any angular direction for completion of circular maneuvering of 360 degrees delimiting extent of discernible dimension and averaging analysis.

i. With a simple drawing EU36, it attempted to clarify how the natural way restricted 'bi-directionally governing' living physique can abridge in revolving manners based on 360 degrees cycle, which satisfies the least multidimensional roles of elementary levels works of energy elements.

(Here, the sense of 'bi-directionally governing' is justified due to monopolizing features of some sensing organs that natural ways formulated by the energy-God following the diversifying physics values roles in a habitable atmosphere/s.)

Multidimensional projection following the structural importance ofintelligent beings

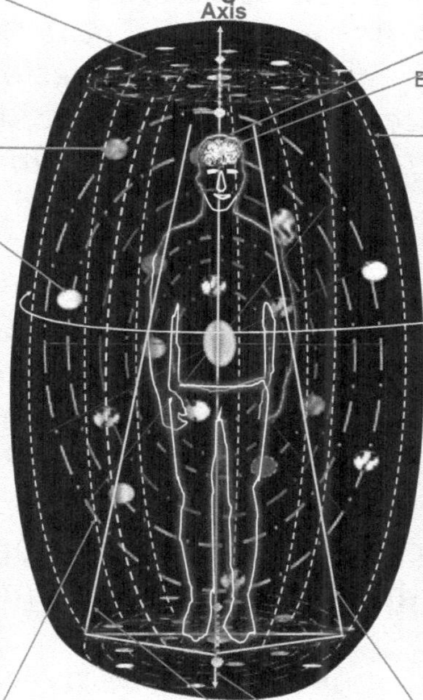

In the drawing, the EU36 indicated some significant points as follows: -

a. Limits of Energy releases for the effectual Physics values derive from the different positioning constituent organs in a zero-status-stable living physique. Remotely the same is comparable with the elemental arrangements of energy elements whenever they unitedly perform as a referenceable zero-status-stable unit-value mass-energy under conducive surroundings.

(Indeed, the physical properties of living organs are not much flexible compared to their in-depth working energy elements when it justifies that an organ is the frame of an overall mass-energy following the diversity of formulaic mc^<2>.)

b. An overall estimation concerning soft-soaping coverage of external (*external means the natural application*) and internal (*means locale administering physics while constitutionally stemming body-Physics derive some form of effective or ineffective energy outcomes*) physics values areas of structurally significant Physical systems of human beings.

→One thing to quote is that the averaging computation does not mean it will give the True outcome at the most-micro momentary levels of variation and any divergence beyond the demarcated boundary of a human being, their group, and even following the analytical output from consciously perceiving natural gift or non-natural way designed aiding that is, types of equipment.

c. So, from the perspective of energy-level activities, talks of reality will remain with the dilemmatic solution as long as the consciousness denies to accept the most-micro levels of work or elementary levels roles of energy elements when the physique of the human being or their constituent molecules are nothing more than the compounded form representing as a zero-status-stable posting mass-energy frame under some conducive surroundings in the balance of internal and external energy-Physics.

→Whatsoever, in this 2nd-energy-belt, the eternity of energies and their diversifying properties are the authentic reality for the consciously granted intelligent beings in this bio-energy resourcing life-orienting inhabitable Globe.

d. Technically, a lively body mass (*bm*) (*regardless of micro to big*) functionally blocks to become unchanging or enduring zero-status-stable position with some amount of energy those are reasonably come to a conclusive status due to internal energy-Physics or the energy-Physics of prospective areas of body mass (bm) and the adequate implication of external-Physics of the surrounding atmosphere to become the mass-energy frame/s referenceable like mc^<2>.

e. Remember, works of the 2nd-energy-belt are in systematic order since everything from imaginable and unimaginable objects, energy, and energy products are interrelated for some purposes.

f. But the deficiencies are the consciousness of the human species and alien intelligent beings and even for their purposes, designed various types of equipment that keep restricting in many ways while attempting to solve vast and tiny levels of analysis and computations, keeping their constituent diversity at a reasonable symmetric status, which justified as a constant or base for analyzation of all others accessible or inaccessible energy and energy-products following comparative strategy in the vast energy-Ocean of the 2nd-energy-belt.

→Also, when physics values are locked by some rules and regulation whenever it represents a Zero-status-stableness positioning mass-energy frame, in concurrence with the surrounding atmospheric circumstances for ensuing the natural way provided rules and regulations of a locale or distant alien locations in coordination with the universal laws of the 2nd-energy-belt, then it can realize through the predominating sensing faculties of bio-intelligent beings and other alien livings that the Zero-status-stable unit-value frame/s of energy-Products, consciously acting with constituent multiple energy-Products that flawlessly maneuvering in the locale or distant location abiding by the universally allocated laws of the energies in the matured 2nd-energy-belt.

→To the active human consciousness, the above also bears a distant resemblance when consciousness appearing initiates from the in-built molecular levels and applies to the entire body mass (bm), which reveals through a different way of internal and subsequent external gesturing while living in a place of habitat.

g. Indeed, the reality of the molecular stage's consciousness may be inaccessible to many intelligent beings when congruous efforts of constituent molecules calmly run the zero-status-stable positioning physical system under encouraging surroundings.

→It also means that there is no intolerable strain on the mental and physically set lively physique from the in-depth union of molecules and the externally (*natural or locale*) or internally (*componential)* regulating physics due to natural and non-natural reasoning, that is, from parametrical viewpoint may say, all sorts of acts and effects are balanced.

→And even though all are in parametric balance, in any case, some damage may happen to a molecular stage that does not allow the brain processor of a human being to analyze and realize to act.

Then someone lively position will remain unaware of the affected molecule/s unless that damage infectiously increases to knock the brain processor (*these micro-level effects are happening to many people, which does not allow them to realize during their daily activities*) since the delimiting brain processor restricts in many ways for assessing direct or indirect to each and individual molecular works of complexly arranged organs in a human being.

h. Besides, from the perspective of energy, the zero-status-stableness of a physical appearance determines by the balancing application of external and internal physics when the physique justifies as a unit-value referenceable mass-energy as a human being and other living.

14.7) Here are a few analogous points about the energy levels' consciousness of three-dimensionally evolved living beings.

Additionally, based on the explanations provided above about the consciousness, it attempts to simplify a little more based on physical reality to someone who accepts that he is not directly aware or witnessed the initiation of his birth until and unless physically evolved and independently realized, as well granted by others as a conscious human being in a circumscribed conducive atmosphere.

a. Upon evolving as a human progeny, bio-globe or shelter-giving bio-mother nature technologically instructs the newborn from the molecular levels and even may think more in-depth at energy levels, how to adapt, grow internal physics (*body physics*), reserving balance of own physical state and can challenge to some limits of external atmospheric applications.

b. Next, progressively from birth, the journey begins to struggle in versatile energy-physics that allows one to acquire knowledge and improve consciousness, even at the molecular level of the living physical union, before the brain processor matures up and can tackle sociological circumstances through the undeniable physical appearance and confining measurable technique as the physical reality of the overall yield of the whole body-mass (**E**$_{bio-energy}$**=mc^<2>**) as a unit-value living being.

c. I think everyone will accept that as a physical state of a living being, a journey of life, that is, the event of life and death is one-tracked (*unlike philosophically imaging long-*

lasting stone sculpture) since that obligation obediently heeded by the predecessors following the technically presented universal laws of energy transformation.

Accordingly, possessing consciousness as a physical state of a living being cannot maintain eternality since it is some class of outcome from a specific type of energy but can keep on mutating.

d. Now, when someone happily and satisfactorily proceeds toward the headway of life as a biological state of a living being, his superficial consciousness suppresses much of molecular-level information, deriving ambiguity when judgment granted by the physique as physical reality and empirical, through the delimiting preset sensing systems and brain processor.

e. Ages of life expectancy are stepping towards augmentation following the base of a parametrically sectioning phase as newborn, childhood, young, middle, and old within the confinement of physical reality as a state of appearance like biological living beings since 'energies of the contemporary standard' are incapable of maintaining any immutable position and give enough supply as the molecular meal of a childhood state and then follow-on succeeding ages.

→Why struggle!

So, natural ways of struggle (*struggle resulting with the exhaustion of old and appearances of new-new molecules*) improve towards the juvenile, middle, and lastly, old age where constituent molecules are incapable of performing the same as the early ages vigorousness due to atypicality of life-supporting energy source during the period of the oldness events as the life of a human being, so all these became a natural mechanism when enumerated based on the three-dimensional pre-set values.

f. So, upon exhaustion of all the events of existence as a living being, only the traditionally believing soul {(*at the energy level, this soul justified as an overall thinner form of energy of the living body mass functions as an animator for liveliness appearance while in existing, besides its regulating quality try to maintain some reasonable balance of body mass (mc^2) wherever resides or travels)*}, act as a witnesser and as well as record everything at the molecular level, instantaneous from the commencement of birth and consistent as physical existence.

g. Thus, the soul is conscious since it registers past activities and prophesies for the future course of the journey of the body mass as appearance living life, stays as the secret holy regulator for bodily appearing human beings in this bio-Globe.

Also, the same can apply to many alien groups of non-bio-intelligent species regardless of their physical attainment, like the noticeable figure as living species in this biological globe or different forms of living mass-energy frame or physique.

h. Technically, at the energy level, 'loss and gain' are consistent practices from a most micro portion to a large amount of energy of the 2nd-energy-belt.

→So, comparably, not of any section from the hugeness holdings of this bio-Globe or any observable and non-observable Universe enables the storing of information and acts consciously like the total energies of the 2nd-energy-belt.

In this regard, it may recollect and say, ostensive Purest-energy is the ‘Whole and Soul’ and perfect regulator of all diversity in the 2nd-energy-belt, whether discernible or not imaginable by human beings or alien species.

h1. One of the most fundamental talks is that the dignity of 'E' energy is eternal, except its essence 'mc^<2>' only follows the diversity and transformation roles, wherever and whenever realizable or not upon induce to natural or non-natural ways appearing physical system.

→In this regard, one thing may requote, from the standpoint of energy, nothing such items or objects exist in the ever turbulent 'energy -Ocean' of the 2nd-energy-belt that can justify as Truly serene.

→Indeed, exceptions are the naturally developed living computing processor such as the brain-box, and non-natural ways designed computing machines for aiding the analyzing and decision-making practices through the delimiting data stored information through the living brain-box, where the most micro to monumental problems tend to summarize or abridge, neglecting the energy levels immense works and consciously conclude at some acceptable position for oneself and the group's necessity as human species.

Chapter 15

Energy and the Gravity.

Basic Concept of Gravity.

The concept of gravity is not the new presentation of the modern scientific era, but the notion and contribution related to gravity began long before. Publicly, in the mythological periods, many scholars from Greek, India, Rome, the Middle East, and Persia presented their ideas and mechanism of gravity.

From the 17th century, many scholars found the presence of gravitational vigorousness, which facilitates follow-up generations' augmented way of presenting the significance of gravity and its pulling effect.

In the modern days two main theories of gravity are most well-known 'inverse-square law' and 'general relativity' choices for observable acting objects, bodies, or masses.

i. Almost everyone in the modern generation is well aware of gravity which is a force of attraction between the units of masses, particles, or objects everywhere it appears. And while doing so, they exchange and transform some form of energy they endeavour, an eternal relation of object and mass-energy frame following the principles of mc^2.

→So, considering the one-way property of attraction, gravity reasonably may term with some form of energy having pulling rights of any unit-value object/s, microparticles, and unit/s of mass-energy frame whenever it endured zero-status-stable positioning regardless of kinetic and potential.

ii. Gravitational force may recognize as gravitational pull or attractive force, exists between all objects with mass. It acknowledged that the strength of gravitational pull depends on the size of the unit's mass /masses energies and the distance between them.

→Due to the specific features of attraction, the gravitational pulls act on many classically presented matters, and even in our daily life, comfortably can access, like falling fruits from the tree, rocks cannot fly, the steadiness of objects, etc.

iii. From the viewpoint of "Quantumology," hypothetically believe that massless elementary particle has their energy, and **graviton** believes to be accountable for carrying the force of gravity.

iv. Following the general observation and acceptance, the word sense force determines the push or pulls of objects, root for deriving acceleration means the rate at which velocity alters with Time, regarding speed and direction. Velocity means some essence of motion or movement for internal origin or external application.

v. One thing to requote for convenience, from the viewpoint of energy, every unit-value mass-energy frame or their union as an object constitutionally developed some form of motion due to centralized motion deriving motion-element of energy.

vi. Considering the above, it may extract and say that whenever there's the presence of 'mass-energy' acting as a framework can resolute its gravity possible of performing independently with the gravitational pulling acts.

15.1) Gravity to Understand.

In general senses, the word 'gravity /pull /gravitational' is considered the universal force of attraction acting between all Matter, including our bio-globe, which relates to the force, motion, and Matter of physical objects (*mechanics*).

i. In the modern scientific era, gravitational theory dominates terrestrial and as-well cosmical phenomena, which conceivably extend beyond classical physics for any mass-energy frame regardless of the imaginable most micro and giant object as a sack-of-energy.

ii. Next is gravity at the quantum level, word sense quantum indicates a discrete quantity of energy definable as a distinct unit of the occurrence, and quantum physics terms as research on the most rudimentary level for uncovering relations of Matter and energy.

And quantum gravity tries to define by unifying the other essential forces of classical physics into a gross term.

Accordingly, the hypothetical particle/s presented with zero charges and rest mass (**E0**), which hold to be responsible /mediating for the quantum of gravitational force.

iii. But, my ideas represent in different ways, which may presume beyond particle physics that, in general, can easily postulate through the standardness of contemporary logic sense.

→Gravity is Truly for energy level/s function before particles (*frames of mass-energy or mc^<2>***) appear with varieties of structural significances (***dimensional aspect***) in this matured 2nd-energy-belt,**

→Also delayed acknowledged many quantum particles that befit to comprehend directly through the standardness of bio-intelligent species in this Globe, besides there are some weird particles which are beyond reachable to specifically designed bio-intelligent system's acceptable indirectly through the standard logical applications, possibility to occupy some low or high degree (*obeying the quality and quantity of innate value's energy*) of gravity generating energy.

iv. Here, I may quote from the earlier volumes clarifying the birth of the Universe and its three independent energy belts from the True singularity status of the Supreme - Universe.

→The 2nd-energy-belt settled with the explicit types of energies and energy components /energy-Elements (ee). This belt is consistent with roughness due to differential roles and distinguishable diversifying properties based on the design and conception of bio-intelligent species.

→In the matured 2nd-energy-belt, any state of unit-value frame of mass-energy (mc^2) qualifies to set structural significance as particle/s and must constitute energy components or energy-Elements with distinctive properties.

→These in-built working essences or energy-Elements are responsible for deriving all internal and external activities, even some degree of pulling effect, which is capable of reserving inner balance /stableness as well the integrity, in concurrences with the surroundings atmospheres (*Twig-atmosphere + energy-Atmosphere*) wherever they are in rest and whenever in the move for action.

v. In particle physics, gravity is one of the phenomena released by the structural significance particle as mass-energy frame due to internal works of energy-Elements having the quality to generate pulling coercion that may consider like quantum way defined '**gravitons**' but dignify them as **energy graviton-elements** since it functions as the 'graviton' the elementary particle.

→By the way, it is to quote that in this modern quantum era, graviton considered as a hypothetical particle with zero charges and rest mass that held to be the quantized gravity, which carries gravity for the gravitational field to explain gravitational phenomena for huge mass-energy frames.

vi. From the viewpoint of energy, one thing is to requote that motion is the root cause for everything regardless of effective and ineffective concerning act and effect.

So, the quality of motion-element and influencing energy-element (*'influence energy element' is one of like other working essences, which can sway over other energy elements*) in a frame of mass-energy or a sack-of-energy as massive structural significance determines the action for others distinctive properties energy-Elements (ee) in the coordination with surrounding/s, and then the gross outcome with typical properties of most-micro as a quantum particle and consequentially to their unitedly shaped huge mass as an object.

vii. Accordingly, if assuming some form of energy-component/s like **energy graviton-element** is responsible for generating gravity. Then it is possible to say gravity is the constitutional part of the mass-energy frame regardless of parametrically distinguishing the lowest and highest measurable limits.

→Likewise, there are many other properties /features /phenomena expressible or not through the distinctive word senses, constitutionally occupied by the frame/s mass-energy that someone may justify as practical or hypothetical in this 2nd-energy-belt.

And briefly say, the constitutionally set energy elements are the working essence for everything, whatsoever bio-intelligent consciousness invents as the outcome and effect of matter, material, objects, and the micro level, which recognize as elementary particles containing some form of 'mass-energy' and representing as frame/s following the formulaic mc^2 or it augmented position as mc^<2>.

viii. Further, conceptional clearance may recite that the word-sense vacuum indicated nothingness /emptiness, which is perfect for the implication of observable items.

→But from the viewpoint of energy, those word senses like the vacuum have no True position since some form of effective or non-effective energies, mass-energy, and most-micro levels particles play all around, whether perceptible or non-perceptible, through the contemporary consciousness of bio-intelligent species following the accessible object in the matured 2nd-energy-belt.

15.2) Herewith elaboration of some significant points about Gravity, as previously introduced in the books 'Mythology, Philosophy, And Science, Book for the New

Generations Volume I and II, presented in a concise and accessible manner.

i. The Supreme - Universe is at State-of-Absoluteness, and its possible units of Pure-energy are at '0absp' and illustrated as: -

► **SU = Supreme-Universe**
= 0absp = State-of-Absoluteness
= Symmetric and bearing ever-constancy on the point of both logically presentable numeric and magnitude.

Or following the above status,

►**SU = Supreme-Universe,**
conveniently considered based on the existing famous equation,

E = mc^2, and may present as

'E = True singularity' that is, (E 'Supreme-Universe')

= E 'Singularity' = E $^{\infty}$ units of Pure-energy),

where there is no representable diversity like mc^2 or mc^<2> which indicates the diversifying activities of energies in the 2nd-energy-belt, maintaining equivalent relation due to genetically bearing essences from **E^{Pure}** of the Supreme - Universe.

a. Accordingly, 'Gravity' is also in cohesion, like the other numerous energy elements at their singular or only one uniform state and their neutral /ineffectual status in any assumable unit-value Pure–energy of the Supreme - Universe.

Based on perceptible sense, from the initiation and gradual development of the embryonic stage for the future

Universe, acts like 'Gravity' comprehensively and effectively seem suppressed.

It is because of some extreme short-burst energy values' instantaneous activities to the units of prospective energies that those who undertook the preparation of the baby Universe consequentially compelled to curb many sensible works and effects that are discernible through the perceivable sense while analyzing the performance of observable and non-observable objects and products of different cosmical-Universes.

b. Further, at the embryonic stage of the Universe, the objectually defined property of gravity, or say, gravitate - value was extremely slow in the aspect of any activity or an objectual sense, maybe assuming following the word as completely idle due to containing units of energies were almost inseparable from the feature of Pure-status.

And accordingly, following the perceptible sense, it can be considered that there is no active form of objectually definable gravity and gravitate-value that can analogically prove the roles of gravity during the formation of the embryonic Universe and continuing its journey before finally locked at some discernible state of intelligent beings' realization in the matured 2nd-energy-belt.

ii. Gradually, from the commencement and then towards the growing process of the embryonic state of the future Universe, due to mounting diversity besides the appearances of various transient energy elements, gravity begins to initiate at the unrealizable weakest form in the occupied areas of the Supreme - Universe for her future clan, the baby Universe.

a. Due to the active roles at the elementary levels of disturbed Pure-energy from the initiatory stage of the prospective Universe, and at some stage following many elementary levels' physics, or elemental values of energy elements lead to cover some attainable areas in the great grandmother Supreme - Universe and consequentially derived towards maturity, constituting three proprietorially different energy belts like 1st-energy-belt, 2nd-energy-belt and the 3rd-energy-belt.

→The 1st-energy-belt and the 3rd-energy-belt independently reached a resemblance of absoluteness, or say, complete neutral state other than their closest barriers that are the acting repulsive-coercion with internal containing of the 2nd-energy-belt.

→Upon the maturity of the 2nd-energy-belt, significant acts of gravity gradually encompass within the limit of 2eb-energy-clock or the comprehensiveness of gravitational action mainly restrained within the 2nd-energy-belt.

→So, two energy belts are the 1st-energy-belt, and the 3rd-energy-belt became the acting base position for assessing the overall energy times for the space occupied by the 2eb-energy-clock.

→Besides, these two energy belts (*the 1st and 3rd-energy-belt*) also stand as computing bases for gravitic, magnetic, and likewise appearing energy impulses during diverse activities in the matured 2nd-energy-belt.

b. Whatever the logical sense applicable to this bio-nature is either recognized as discovered or obscured about the true meaning of Gravity.

But for the instance of energy in the matured 2nd-energy-belt, Gravity is certain kinds of releases and posteriorly /latterly developed objects, objectual products. And like the advanced stages' matter-energies, matter-elements, matter-products, and so on products due to diversified roles of mass-energy when represented as some form of frame/s.

(Drawing EU36a A simple view of 3D Projection of Gravity.)

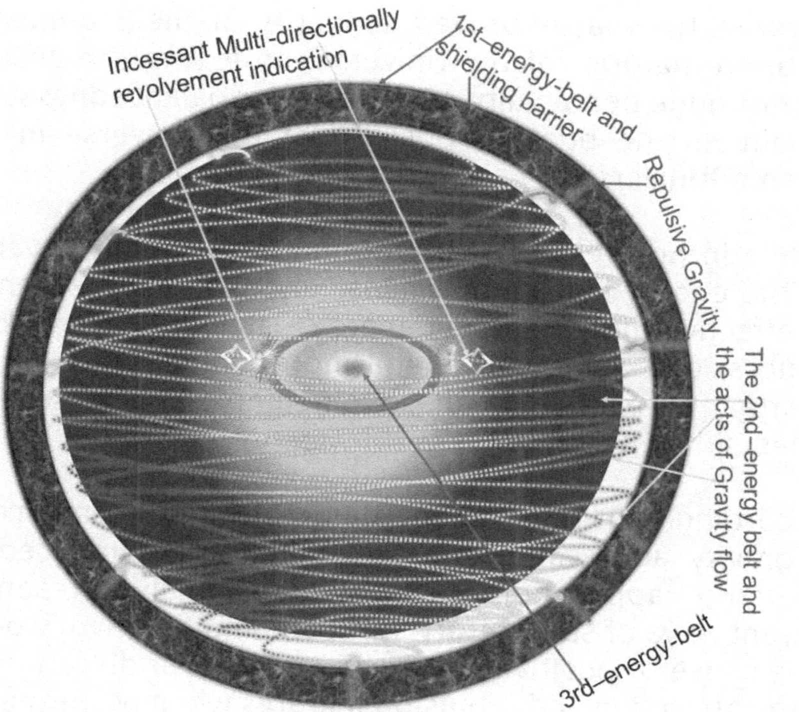

The energy levels Pulling force all around the 2nd-energy-belt, which are consistently resist by the 1st-energy-belt and the 3rd-energy-belt.

EU36a (A simple view of 3D Projection of Gravity.)

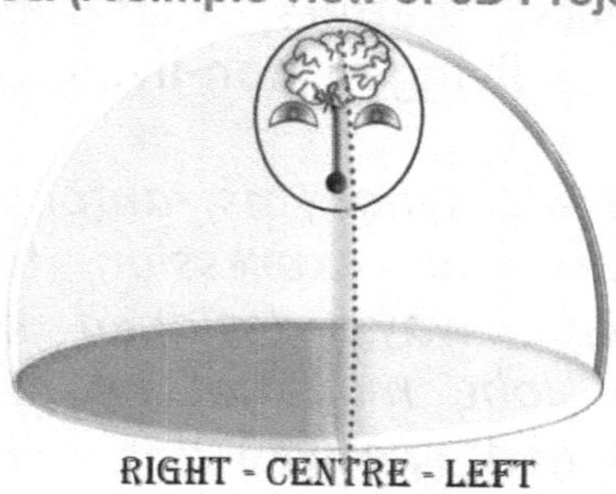

iii. By the way, gradually toward the development and maturity of the 1st-energy-belt, it filtered off, holding many significant ingredients as energy - elements before discharge or push-off residues in the 2nd-energy-belt.

And then, incremental filtration and derivement of the utmost thinner form of energy or Purest-Energy, which is characteristically extreme to highest of dynamism and neutral status that improved solidarity of the 3rd-energy-belt upon maturity of the Universe, so it (*the 3rd-energy-belt*) got huge personality and rules over balancedness and to maintain the buoyant position of the Universe in the Supreme -Universe.

a. While in progress towards the maturity of the Universe, the 2nd-energy-belt comprehensively lost many essential and effective ingredients as energy - elements or working essences that contrived them more unbalanced at the grassroot of energy means, elementary level of energy elements.

→Accordingly, in the 2nd-energy-belt, the performance of the gravity aggravated and substantially commenced to work and application through objectually presented different ways of satisfactory synonym and antonym words, or the sense of positive and negative ways' of discernment like the attraction and repulsion features when performance delivers following the findings of the frame of micro to massive unit-value light, dense, or compact state energies, representative objects, matters, materials and so on observably consider as living or non-living products.

(Here may note, all the synonyms, antonyms, and many other words utilized for expression that befits the realization of specific ways designed modern human species in this bio-Globe may not be apprehensible to advanced alien intelligent species.)

→Although the science of this biological world may find some relevant conceptions about the acts and effects of gravity based on the available physics -values act on various states -of- energy, recognized as solid, liquid, gaseous, and plasma.

Nevertheless, the act of gravity and affectivity of many kinds of elemental values of energy elements in the neonate

2nd-energy-belt is remote to analogous with the objectually bearing characteristics of gravity. Since releases, as gravitate energies from unit-value mass-energy frames, microparticles, objects, and types of energy products are the multiple derivements, consequences of internal and external energy levels' work.

b. In this matured 2nd-energy-belt, generally, the mensurable gravity and its potential works of lately developed objects or mass-energy frames are due to the development of specific forms of internal energy (here, specific form means not primordial energies, but the following stages evolved in particular from incongruous types) and as a result of releases from the formulated componential energy elements' functions.

→So, the root cause of gravity may consider some forms of energy impulses /signals released due to internal acts of energy elements to any mass-state energy as a frame, or a unit-value sack-of-energy represents an object, and those impulses capable of performing gravity-related works multi-directionally, including particular choice as positive (pulling /attraction force) or negative (pushing /repulsive impact) ways of discernible roles.

iv. By the way, it may recollect, from the viewpoint of energy in this matured 2nd-energy-belt, gravity is possible to assess as unit/s of the tiny and massive state of mass energy pre and post state's frame, based on the 2eb-energy-clock that is the holding energies of the 2nd-energy-belt and the mensurable energy-Times of 2eb-energy-clock.

(Here, the 'pre' stage's mass-energy means a primary or immature frame of mass-energy and not yet prepared to act structurally representable object or particle, and the 'post' stage's mass-energy frame means the mature status of the mass-energy frame and is ready to represent an object.)

v. Here, briefly, on the point of energy, it is to note that the root cause of Gravity relies on the Motional-activity of the motion element. The Motional activity is accountable to the aggregated quality and quantity of containing unit-value energies, which are supported to structure an objectual-element like matter-elements, material-elements, materialistic-elements, and thereon at developed stages consolidation as large matter, material, and materialistic product or object.

→Now, on the point of 'gravity' or the attraction force, if there is no sense of word or application of motion or motion-deriving element like the formulaic way presented symbol 'c' of (mc). Then, the question of 'gravity' and its related working items may not arise.

And accordingly, it also will be invalidated to say that some things constitutionally concealed in unit-value energy, object, or union as a mass-energy frame, irrespective of any form of defining physics values like actions and reactions due to pressure, force, temperature, influence, gravitational works, etc., will be the root cause of all activity and inactive roles we observed in daily life for the acknowledged matter-products like articles, particle, etc., and so, when something is at rest or in the move will be an everlastingly consistent feature.

Hence, the word sense mystery will dominate for all direct and indirect computations or analyzations when the conclusion performs following only the averaging three-dimensionally skilled intelligent beings in this biological world.

→Therefore, it can justify, in a unit-value 'sack-of-energy' besides many types of energy elements, the motion element became the root cause of deriving motion. Likewise, the constitutionally gravity-generating energy element is more mighty than other energy elements in concurrence with the surrounding Atmosphere {*that is, the*

Twig + (Energy-atmosphere)} performs the role of gravity for any form of an object/s. Irrespective of a large and micro mass of energy like our globe or a living molecule.

→So, based on bio-intelligents' logical sense and following the diversity of constituent and diversifying intramural works of energy-element in a bio-mass or biological energy-sack, it may justify that all sorts of the units as objects or products form, more or less competent to generate a certain degree of gravity for their internal purposes.

vi. The naturally borne word 'gravity' and its related essences, regardless of ideologically presentable gravity and Anti-gravity, are the living ethics of positive and negative senses. And those senses befitted to some dimensionally secured conceptions of bio-intelligent species, which logically contributed to satisfy the practical and hypothetical items and all accessibility structural diversity reasonably to understand in this 2nd-energy-belt, including observable, non-observable things, and multiple transitionally appearing items in this 2nd-energy-belt.

a. Accordingly, the concept of Anti-gravity, in general, is considered free from gravitative coercion for the acts of objects.

But technically, from the perspective of energy, the word sense 'free' and 'freeness' is inappropriate. Instead, some classes of energies, from the low-grade to high-grade, are all-around and always ready to accept an action or react whenever an object or mass-energy frame performs with something following the universal laws of diversity in the 2nd-energy-belt besides natural ways of abiding transformation rules.

→So, the sense of the eternal role of an object, matter, material, and other products never happens. Instead, the word sense everlasting /eternal is a relative satisfaction to bio-intelligent species and the judgment through inbuilt delimiting measuring system directly or aiding with devices like computers, Artificial intelligence, etc., comparing to the interrelating roles of non-fathomable discordant types of energies and energy-Products of energy-Ocean in the 2nd-energy-belt.

vii. Following the general observation, the conception of gravity is a latent infix /inherent feature of an object /mass-energy frame (mc^2), like objects, matter, materials, and so on that are recurrently accessible or not to our anchored sensory systems.

All the developed and latent infix /inherent features of an object, like the essences of gravity acquired genetically from the Pure-energies of the Supreme - Universe, a status of **True singularity** (***E Supreme-Universe = E Singularity = E ∞ units of Pure-energy***), securing all sorts of equivalent relation (~) in the form of mass-energy, energy-products or objects, besides the inertness of all imaginable and unimaginable diversifying capabilities in the Universe.

→So, due to the inherent energy level's property of 'Pure-energy' of the Supreme - Universe, any form of mass-energy frame and structurally discernible object in the 2nd-energy-belt capable of acts related to gravity and many other physics values roles, which consistently repositions between the energy sources whatsoever observable products (*that has structural or some dimensionally discernible values*) inventively appear with new-new features and properties to bio-intelligent beings.

viii. Objectually presented Gravitational force is a one-sided conception that is an attraction force and observably applies as some form of directional physics have the pulling roles only.

Based on simple sense, it may assume that everyone is aware that the 'Gravitational force' recognizes since it acts as a pulling role between the two bodies, directly related to the beholding or integrally occupying masses, and inversely proportional to the square of the distance between the objects.

a. But in the viewpoint of energy, the sense of gravity may conveniently dilate into direct and indirect ways, utilizing synonymous and antonym expressions and implications like the attraction and repulsion of physics values directional activities for convenience to grasp many known and mysterious outcomes as a result of gravitation-coercion of objects /mass-energy products perceptible through some delimiting dimensionally projected natural and non-natural items or equipment.

ix. Justifying the logicality, objectually recognizing anti-gravity or repulsive-coercion or repulsion due to gravitating coercion at energy levels, anti-gravity deters many unsupportive /harmful turbulent actions from the 2nd-energy-belt that may affect the progression of the 1st-energy-belt and the 3rd-energy-belt.

Also, the same may apply to delineate that upon maturity of the 1st-energy-belt and the 3rd-energy-belt. All physics values' implication, including conceptually developed overall anti-gravity-coercion, blocked and effectively isolated at the individual shielding barriers with the 2nd-energy-belt, retaining intact all the diversifying roles of energies in the matured 2nd-energy-belt.

x. At this moment, in an objectual way, justify that the cause of generating anti-gravity impulses is due to directionally differentiable activities among the energies and energy products. Besides, colossal differential - values activities in the matured 2nd-energy-belt, so the greatness of anti-gravitate actions to bring back some state of normalization.

→Here, the conception of susceptible /sensible gravity mainly relies on the features of matter products that later developed due to the coagulation of multiple tiers of advanced stages' energies.

→Accordingly, the word-senses appearance and disappearance of objects and objectual products get validated concerning their existence following the transformation rules.

→By the way, it is to say, all forms of objects and objectual products strictly observe the rules of transformation with the assistance or in coordination with the physics-values activities like gravity, anti-gravity, and so on, numerous discernible and inappreciable roles.

xi. In the 2nd-energy-belt other than the thinnest form of energies, any frame of unit-value-energy should occupy at least some minimum values energy-elements, besides inseparable /non-isolable motion-element capable of inciting differential-value, and efficacious generic gifted feature as gravity and likewise internally arising (*that is, elementary levels of energy elements*) energy-impulses.

→According to the specificity of bio-intellect design, gravity on the sense of attraction or attractive-coercion, there should be some proportionate form of anti-value or deterrent coercion either to a level of low or high negative value or may justify as Anti-Gravity-coercion or repulsion.

Persistency of gravity with the dignity of attractive coercion and repulsive coercion, or is justified as gravity and anti-gravity, are due to the inseparableness degree of momentum to the containing motion-element in a frame of unit-value energy.

The appearance of mass form energy-impulse, or the gravity-forming energies, is perceptible in enumerable descendant sub-classes energies, even to the living and non-living objectual products.

xii. During the progression of the 2nd-energy-belt, besides many other efficacious physics values works, the attraction of gravity and the repulsion of gravity (*that is, gravity and anti-gravity)* is treated to drive out the fine and comparable highly differentiable and most thin /diluted state of energies, like motion - elements that effectively works out /formulate for the preparation and maturation of the 3rd-energy-belt.

In this nature, objectually presented gravity usually assesses conceptionally limiting times and areas integrally set in bio-intelligent computing systems.

Accordingly, gravity is relative throughout the matured 2nd-energy-belt to the finite of discernible limits, or the boundary of rational sense to the standard of positive and the negative, or say, as the directionally defining measurable tool like a specific values' parameter or measuring frames.

15.3) Why is gravity so significant for our planet?

Now, if someone assumes no sense of 'gravity' applicable to the Earth as a mass-energy frame and the containing objects, irrespective of living and non-living items. Then, from the viewpoint of energy, the root cause of gravity generating coercions or gravitating-coercion (*or gravitational force*) due to the motion or movements of the Earth will not be applicable.

And if the Earth stops movement, everything related to the Earth will be imbalanced from the originality determined following the universal laws of the 2nd-energy-belt.

i. Some points for elucidation, in common, acknowledges that the Earth has two types of motions 1st one is rotation /spinning on its axis, and the other is a revolution around Sun in a fixed orbit.

→The word sense gravity conveniently may consider as a pulling effect, and as a result of the pulling effect, compositionally bearing apparent *(mean not everlasting)* stable properties of items are displayed with proper types of motion such as rotation and revolution in the Earth.

→So, the sense of gravity means attraction /pulling effects that tend to hold things firmly, closer to gravity-generating sources (*here, the reference 'source' may justify as the energy levels' motion deriving motion elements*) like this Globe.

→Now, compositionally, assume that the Earth is a massive mass-energy frame alternative to the multiple types of properties of microparticles, and at the quantum level, all particles may not qualify to generate gravity.

→Accordingly, the overall outcome of the Earth's 'mass energy' determines the pulling effect to free-flowing objects or particles nearer and within the limits /space of the gravitational field.

ii. Stopping the movement or spinning of the Earth will not mean the total ceasing of generating gravity at the microparticle level. But dispersal and creation of gravitational field will be narrowed down /collapse in addition to many side effects to the occupants of the upper layers of the Globe.

→Also, the internal compositae of the globes will get affected, which may lead to total disbalance from rotational and revolutionary movements, and will gradually lead toward destruction from originality.

→According to the excessive centralized pulling effect of gravity, the compositional mass energy, in general, will not allow it to launch any inhabitant into space. Instead, it will create a strong pull centrally, so all the body masses (bm) of living physiques will tend to compact on the point of dimensional significance.

→For example, someone may imagine a moment about the formation of the Sun from gas and dust particles or a nebula for collapsing of gravity.

→So, it is easy to guess what will happen to the Earth when stopping or disbalance of movements totally or partially.

→The above ideas are primary presentations only, but from the fictional point of view, immediately upon proceeding toward disbalance, the Globe will initiate losing self-deterministic control (*self-deterministic control is naturally formulated following vast energy levels' works of the energy-Ocean*), resulting in a call for changes of its many dwellers, including the human species following the laws of survival in this bio-nature.

→Accordingly, changes from the micromolecular levels may lead to a total deformation of the most physical features of specifically designed living species.

iii. Now think, what will happen, if there are no gravitational activities on Earth?

In this regard, I will go for a funny fictional answer someone may enjoy beyond the purview of quantum ideology.

→Any positional disruption of movement, regardless of spinning and revolution, will affect the Globe and all living and non-living dwellers in it.

→Anyone can imagine that upon ceasing of rotational movement of the Globe, naturally, it disturbed one of the main parametric tools for balance retaining its steady orbital move.

→So, gradual process, there will be chances for dimensional defacement of the Globe from the present status due to massive differences in appearance between daylight Times and night.

→Therefore, from the observational viewpoint, the floppy mass of the body of the Earth may turn flat or may get other shapes based on compositionally existing features of mass-energy frames, objects, or matters in their diverse state like liquids, gases, solids, and other forms of particles in different locations.

→Besides, following the survival or the act of inhabiting necessity or reserving a Zero-status-stable positioning of the living and non-living physiques, and from the constituent unit-value frames of the mass-energy of the dwellers in the Globe will get deface from originality.

→Accordingly, the structural values of many lively species may transmogrify to cartoon figures that many of the old and youthful modern generations visually enjoy or play in video games.

→For more examples, think of the olden days when pot makers employed flywheels for rotation around the axis to give the desired shapes of pots. Potters manually controlled the rotations /spinning of the wheel according to the developed consciousness he gained through experiences.

→Here, to ease of understanding, a few technical points to define that when the flywheel spins, the natural way it generates a gravitational field to some limits that control the molecular stage of soil, spins to shape the pot in concurrence with the surrounding physics values.

→It is easy to comprehend if the control of spinning of the flywheel is in excess or less of the parametric value that gives a specific shape. The projected structure of the pot will be affected by deep molecular levels depending on the applying rate of rotations besides many known, unknown physics values actions of the surrounding that natural way requires to give possible perfectness for any specific choices pot.

iv. Now assume there's no orbital momentum of the Globe to any extent.

Accordingly, self-axis momentum will lose total control initiated from the microparticle levels, resulting in regular bonding effects of diversified energy products as objects.

→One thing is logical to say united efforts from the most microparticle levels of appropriate energy products gave the reasonable shapes of objects that bio-intelligent species consciously accept and enjoy beholding oneself as a constant value for bodily appearance and judgment of everything access.

→Based on contemporary logic sense, upon total loss of orbital momentums, initial self-axis rotation of the Globe will increase excess (*or if someone does not believe in 'increase' then, may consider as declination disturbing the standardness of spin, following its' own axis from north to south*) to the limits of the parameter that compel to break level-headed bonds in particle statuses. Thereon, consequential surrounding physics values impact will lead from deforming to destruction.

v. Superiority or 'Prideness' loss of intellectuality of bio-intelligent beings!

Now imagine someone is observing from a distant planetary system the disastrous situation of this bio-Globe and the bio-intelligent species. Indeed, compassionately, they will feel extreme regret for the miserable circumstances of the human species in this Globe since relinquishing the long way struggle and acquiring consciousness that dignified them with pride as intellectual in concurrence with their surroundings.

→Further, due to improper rotational disbalance from the standard parameter, which is gradually set up in the system, or says, developed due to evolutionary process from the molecular stages of ancestors and somehow managed to carry forward from generation to generation.

→Foremost, rhythmically evolved mass-energy frames acting as molecules will lose their **fine-tuned** set up of constituent components or bio-energy elements from living-grade energy, so the driving energy-Times of the energy clock as a bio-molecule will generate wrong and disturbing energy signals from a befitting standard for the whole physique.

It may mean that it will consequentially disturb the level of a mass-energy frame of molecules, and the choices of energy Signals of interest will have no correlation and convolution.

→So, the resulting outcome of the biological clock from the independent molecule will mismatch rhythm that allows developing bio-processor placed in the brain box of an intelligent physique.

→Upon mismatch of rhythm from compositionally deriving informative signals will derive imbalance and make unrest the physical system from its grassroots level of a physique.

→Loss of information from molecular levels results in desultory with new body physics that will tend to disturb regular requisites chemical processing and lead towards struggle and toiling to survive from the standardness of source.

→So, step by step, the mass-energy frames or molecules of the bio-physique will begin to break down from all sorts of internal communication that intercept now and then, hence abating propellant zero-status-stable position of all units and the physique as an acting union human being.

→The sense of fine-tuning.

The sense of tune means adjusting or adapting (*something*) to a particular purpose or situation to achieve the best or desired performance. And, most of the living is finely tuned following the adaption process (*not reality at one go*) in a place of locale for survival necessity in a conducive physics values surrounding.

Further, the above temporal idea is Truly not applicable to the 2nd-energy-belt. Indeed, tuning and fine-tuning happen regularly whenever some amount of energy attempts for sublimation or transmogrification into mass-energy frames and then different analyzable structural values following some dimensionally detectable temporal objects.

Also, obeying the universal laws of the energy in the 2nd-energy-belt, everything automatically gets tuned to reserve the zero-status-stableness position by applying elemental values (*that is, internal physics values*) of a frame of mass-energy or an object, following the regulations of eternal diversity.

The temporally positioned intelligent species always tend to obey the natural and non-natural technologies in balancing with the surroundings' long-lasting diversifying events, so the physicality natural way tries to cultivate and fine-tune internally, augmenting with some forms of defensive and attacking techniques in anticipation of sustenance and survival with dignity in a place of conduciveness.

15.4) Gravitational coercion acts deep in the molecular level of any bio-physique. (*Drawing EU37 and EU37i analogously projected with an 'energy-Clock' and a*

standard clock, presuming them the representative of energy as a sack-of-energy or mass-energy frame.)

Here, projecting with sketches (EU37) to convince least ways that any unit-value bio-molecule or their union as a giant animal, regardless of their distinctive functional organs, can be present as an aggregated packet or sack of bio-energy, constitutionally set-up multiple bio-properties energy-elements (ee) for all functionality of a small or big physique.

→From the constitutional viewpoint, every possible micro molecule acts as an independent 'sack-of-energy' even though compositionally internally connected or intertangled for the overall outcome of the giant physique they united.

→Accountable 'energy-Times' (drawing EU37) can consider delivering from the 'energy-Clock' of the independent molecule. But for the union of molecules as a giant physique, the overall outcome of the molecules, or multiple bio-energy Clocks' multidimensionally providing Times, will represent the overall Time of an entire bio-physical clock of the integrated molecular union.

In a bodily system, perfect bio-energy Times derive following the standardness of internal and external regulating atmospheres from deep inside the molecular level, or elemental level of unit-value energy of bio-energy elements.

→At this moment concerning gravity, say bodily systems connected to earthly deriving gravity, so the Times of bio-Clock from molecular level effect due to acting roles of constitution, set-up bio-energy elements deep in the molecules.

Therefore, variations of gravitational coercions of the surrounding atmosphere, whether by the earth or from space, will act in depth at the molecular level, disrupting the systematic bio-times of the whole physical system, in general, which set up in a system gradually in coordination with the surrounding Times of residences or a native place by birth.

→In the drawing EU29 shown, the centrally situated motion element of an aggregating packet or sack of energy assumes some class of bio-energy representing a human physique.

→Whenever any unit-value energy reaches a conclusive position for representing a zero-status-stable positioning in a conducive atmosphere, the natural Time clocks, in coordination with the supporting area/s, organize the constituent energy-elements of the unit-value energy and thereon union as a sack-of-energy of a whole-body system.

→ In any way, in drawing EU37, some indicated points to understand the leastways about works of energy-Clock.

i. A sectional view of a frame of mass-energy representing an 'energy-Clock' due to averaging multi-directionally deriving elemental-Times of energy elements, attempting to show a reasonable dimension that befits to comprehend by the intelligent beings following circular way revolving and three-dimensional discernment of observable items as machine-made clock's hands act as 2-D rotation, sticking the central position that represents as a 0-D constant for calculating Times of overall feed from different sections in addition to the countable Time indicates by mean of rotating hands or digital methods.

ii. Overall, directional feed from different directions derives due to the consistent functions of energy elements and derivement of elemental values, or physics values.

iii. Energy elements consistently deriving waves of energy signals or elemental-Times in concurrence with internal and external physics values activities.

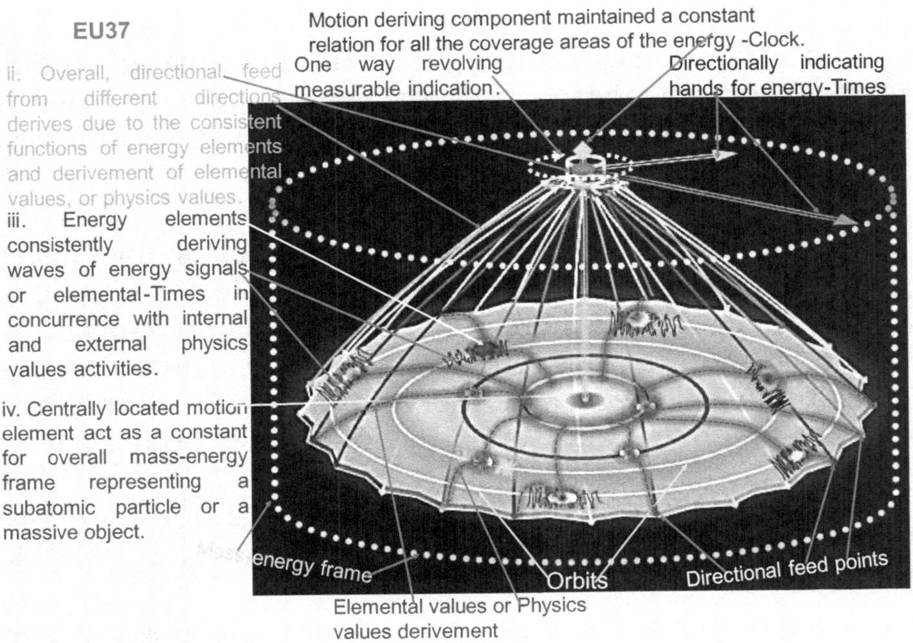

i. A sectional view of a frame of mass-energy representing an 'energy-Clock' due to averaging multi-directionally deriving elemental-Times of energy elements attempting to show a reasonable dimension that befits to comprehend by the intelligent beings following circular way revolving and three-dimensional discernment of observable items, as machine-made clock's hands act as 2-D rotation, sticking the central position that represents as a 0-D constant for calculating Times of overall feed from different sections in addition to the countable Time indicates by mean of rotating hands or digital methods.

iv. Centrally located motion element act as a constant for overall mass-energy frame representing a subatomic particle or a massive object.

→Further, for analogical understanding, the EU37i drawing depicts some sections of a standard clock that display countable times.

And, based on the standard clock, presenting a comprehensive idea to analogous involving energy components (that is, energy elements) deriving overall energy levels Times, which feed directionally to discern conveniently to match the specific way designed human species, keeping a center position as constant for directional circulations on measurable indicator like a dial.

The drawing EU37i depicts a Machine-made clock's hands acting as 2-D spinning, adhering to the central position tool represented as a 0-D constant for discerning one-way measurable Times.

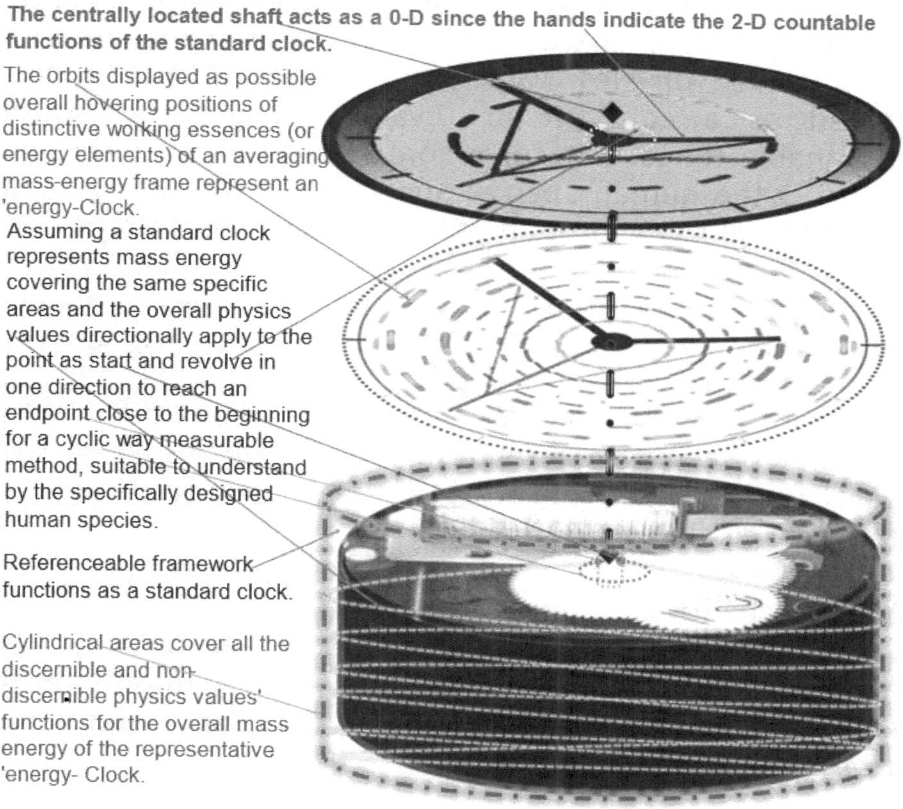

For the standard clocks, the word-sense harmonic oscillating methods, either through mechanically swinging pendulum or applying solar, electrical, and battery power for almost all timekeeping mechanisms. Besides that, following different parts that generally controlled counter chain actions to display accurate time onto the displaying counter.

The concept of 'energy Clock' employs somehow the same approach as applied to the standard clock, which follows the harmonic oscillation as resonance, using some components that vibrate at a natural resonant frequency, depending only on its physical characteristics and being able to resist vibration at some reasonable parametric rates.

But for a typical clock, the operating and non-operating sections concurrently work as a mass-energy frame. So, the standard clock or 'energy-Clock' may consider encountering energy-level physics value applications, parametrically locked as a constant or 0-value, to get the overall reasonable times that solve some purposes of accessible items to the human species specialize of analyzing some dimensional roles through the preset physical system.

So, in the case of the standard countable times of a clock compared to energy-Clock, there are tremendous unimaginable variations from the elementally deriving Times of energy elements and then follow the overall outcome of a 'mass-energy frame' that works as an energy-Clock behind the observable times delivered by the standard clock.

15.5) Herewith are some technical aspects of the physics values that regulate some amounts of 'energy' required for forming a mass-energy unit and then advanced stages particles in harmony with the reasonable atmospheric applications.

Technically, one thing to remember following the perspective of energies whenever some amount of 'energies' transform for the pre-preparation into a mass form energy due to the conduciveness of external atmospheric roles (*that is, physics derived due to the activity of energy-Atmosphere and Twig-atmospheres*) while the compelling areas of the prospective energy for forming mass-energy, gradually initiated to creates and organizes the energy elements as the preparedness of constitutional set-up to become self-dependent and to maintain balance of its structural cognizance in concurrence with the reasonable external atmospheres.

It is the starting point of a baby mass-energy frame for further developments while encountering consistent physics values diversifying activities, following eternity of transformation roles either succeed or fail while facing ongoing evolution and devaluation to take recognizable quantum particles.

Successively, they unitedly positioned small and massive objects with multifarious physics values performance, including externally disposable gravity and gravity-related functions like gravitational field, gravitational force field, etc.

So, in a summarizing manner, all terms of physics derived due to energy level diversity and its productive values are conveniently determinable when some amount of energy transforms to a reasonable structural position, directly or indirectly perceivable through the consciousness of human beings.

From the standpoint of energy, the design of human beings in this mother nature bio-globe is always more exceptional than many others since humans occupy a high-quality regulative processor, like a brain-box, to determine the best control of internal physics values activities derived by the integral molecules in harmony with the reasonable external atmospheric physics values exercises.

15.6) Comparable explanation about the word sense Gravity, Gravitational field, Force field, Orbit, Orbital lines, Orbital-Field areas and their utilization at the elementary level of energy elements, Elemental-values, Physics-values, or elemental-Fields of energy elements in a zero-status-stable unit-value mass-energy frame, and energy-Field of the same mass-energy frame.

To better understand the classic concepts related to some categorized states of matter as solid, liquid, gaseous, and plasm, and concerning some of their functioning terms such as Gravity, Gravitational field, orbit, and orbitals, it may be helpful to delve deeper into their elemental level's functions of energy considering individual forms of unit-value matter having distinguishable properties since occupying some amount of mass-energy that usually approves and accepts through the developed intelligent living consciousness.

i. ► Gravity: -

In classical physics, the term Gravity acknowledges as a force of attraction working between all forms of masses acting as objects.

→In particle physics, hypothetically believes that Graviton is considered a massless particle, which is the root cause of generating the 'attraction force' like gravitational attraction.

→In brief, from the viewpoint of energy, Gravity may be justified as a constitutional property of a zero-status-stable mass-energy frame (*here, the zero-status-stable position means a state of the advanced unit-value mass-energy, ready to transmute or desublimate to a structural figure like the tiniest particle upon application of appropriate external physics values*) and can be defined as the outcome exposed from the periphery of a unit-value structurally substantial object representing the referenceable zero-status-stable mass-energy frame due to the role of qualifying energy elements like Graviton of a typical mass-energy frame/s.

→So, Gravity is also something similar, that an object of massive compact state mass-energy frame exposed with distinctive functional properties like Temperature, Pressure, etc., due to internal elementary levels physics values act while justifying that the object is nothing more than a compounded mass-energy frame bearing the dignity of famous equation E = mc^2 or its augmented position as mc^<2>, means a standard level of Energy 'E' maintained equivalent relations with its constituent multi-grade mass (m) and an averaging velocity deriving source (c) or motion-element surrounded by specific areas *(here specific areas may mean as some measurable extent of a distinct kind)* like ^<2>.

Accordingly, following the perspective of energy, even at some point of imaginable space fabric, there is the possibility of deriving some level of determinable Gravity.

►Some analytical notes: -

From the standpoint of energy, the word sense massless is apparent and inappropriate since Energies are everywhere in the 2nd-energy-belt, whether ply as waveforms, appear as mass energy, or referenceable mass-energy frames recognizable most-micro particles or their compounded way positioned multiple types of objects.

So, I think the sense of massless (*no mass*) subatomic particles or with-mass of atomic particles can represent the unit-value mass-energy frames, referenceable with some discernible identities and properties like most micro-level objects or structurally understandable items.

→Besides, it may requote that the original state of energies in the 2nd-energy-belt is immensely diverse, except for the thinnest monostate (*or something like singleton*) form of ostensive Purest-Energy in the matured 2nd-energy-belt.

→So, it is next to impossible to determine based on typical longest to shortest wavelength releases by the objectual /accessible items through any measurable parameter, justifying them as truly energy level's wave spectrums.

→A little more may clarify concerning the zero-status-stable unit-value mass-energy frame.

In a remote sense, following the spectrum of elementary level of 'energy elements' when a tiniest unit-value mass-energy frame either of lowest to a parametric value's quality, and also the highest with the minimum positioning of two energy elements, and further presuming many more to some uncountable position.

Even then, there must be the most essential 'motion element' and then supposition of the next stage's minimum of one and possibly more and more other energy elements, **sequentially** following some standard proprietorial or far and near denser degrees (*clarified in earlier books*).

→The above **sequential** arrangement is somewhat like the influencing 'energy element' a little denser in degree, orbitally positioned next to the motion element in a unit-value mass-energy frame, and then stepwise many other higher extents which have **versatile and interchanging** properties following the external energy-physics in a conducive *(or parametrically determinable conducive status)* surrounding.

Here, **'versatile and interchanging'** properties may exemplify the linguistically discerning individual alphabets that systematically set from the letter a to z, but their different placements while making sensible words. Also, it can see the planning of many tournaments, players select for their particular purpose of action. However, when the game is on, obeying the winning strategy of the game and for offense and defence of action, players keep changing their pre-selected placements. *(In this regard, analogically presented in the next based on a football game.)*

→[{(By the way, remembering from earlier books may requote, in a mass-energy frame, orbitally positioned **Influencing-element** justified as next to the motion element following the universally standard differentially stationing innumerable energy elements, where the proprietorially thinnest motion element acts as an absolute position like the mathematical figure '0' and succeeding accepted figure '1' represents as **Influencing-element,** and so on stepwise more elevated to some limits of modern intellectual conception on the point of recognizable quality *(drawing EU38, is a standard model on that topic*).

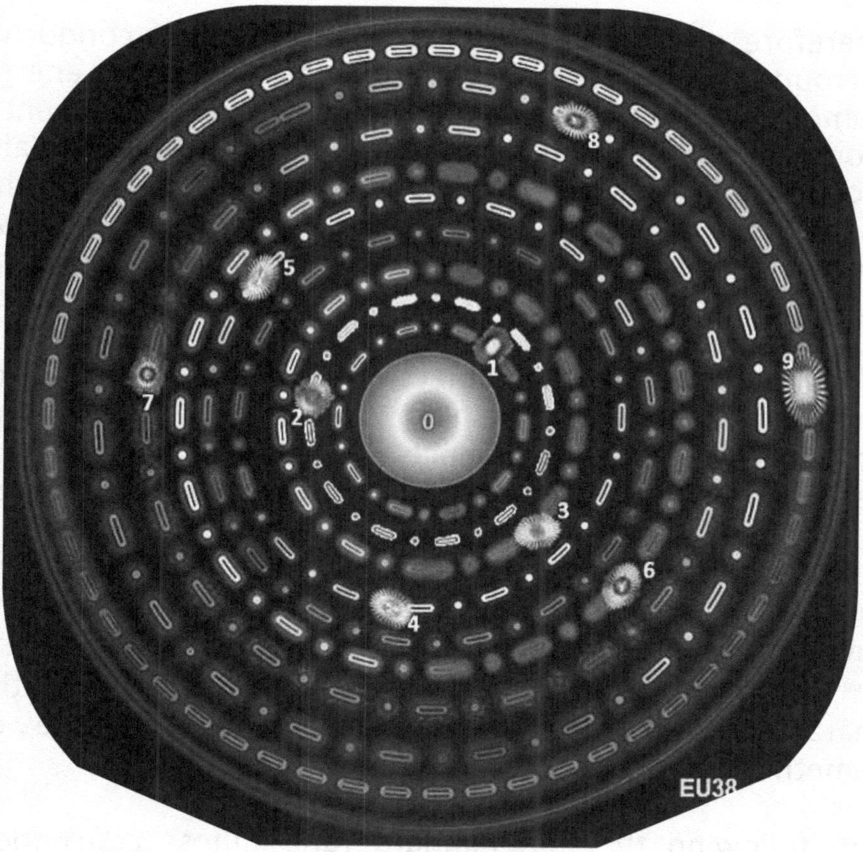

Otherwise, for non-standard ideas based on extreme randomness, any energy elements (*other than the absolute positioning '0' the nothingness or 'c, the behavioral property of motion element as motion only*) bear versatilities (*here versatilities may mean changing a standard elemental position, that is, altering assignment on the aspects of*

degree, act -and- effect due to consistent variations of internal and external physics values roles) to reserve constitutional validity in the ever-turbulent ocean of energy, following the gift (*here the sense 'gift' may represent the tendency to secure zero-status-stableness, something remote of Pure-energies that secured state-of-Absoluteness of the Supreme - Universe*) of great grandmother the Supreme - Universe.

Therefore, upon variation of external physics in a conducive surrounding, any zero-status-stable unit-value mass-energy frame, either with two or more energy elements, momentarily keeps on changing their proprietorially recognizable perceiving energy releases through the assessable objects (*or that object/s may say as* ***full-fledged*** *mass-energy frames having alterable property upon the immediate instant of energy-Times due to diversified sequels of physics values activities)* like the functional property of an object with generating essence of parametrically distinguishing objectually emanating products such as Influence, Pressure, Temperature, Resistive, Gravity, etc^n on the point of assessable effects.

It also means a standard parametric status of an assuming tiny mass-energy frame or particle that proprietorially acts similar to an influencing energy element $\{\mathcal{E}^{Influence} = (mc^\wedge<2>)^{Influence}\}$ having the capacity to sway and can serve on the character, development, or behavior of someone or something, and as well can affect own self.

But, following the non-standard randomness conception means the changeability status due to consequences of diversified physics of internal and external implication, constitutional property of '$\mathcal{E}^{Influence} = (mc^\wedge<2>)^{Influence}$' will alter instantaneously to a new one.

So, the retainment of some undiscoverable curious question of non-parametric measurements about the exact property of free-flowing or the no-mass state energy and all inaccessible mass-energy frames following some determinable in-built conception of many living species.

By the way, one thing may requote that perceptible acts and effects of an object, whether are the productive and non-effective outcomes derived due to the works of integrated energy elements, that is, the solo non-ceasing functioning essences are eternal to an object when justifying it (*objects or subatomic particles*) as a mass-energy frame with mass or without mass (E0).

(*One thing to recite is that rest-mass, no-mass, or E0 status may present through structural value, depictable as a mass-energy frame or an object like a quantum particle.)*

From the perspective of quantum particles (*that is, the zero-status-stable representing mass-energy frames*) with a parametrically determinable property, assuming an alphabetic capital 'P' in a locale atmosphere, which (*that is,* 'P'*)* can relate to the distant particles proprietorially more or less the same as the vocal pronunciation and utilization of other upper and lower-case's alphabetic fonts (**P**, **p**, P, p, *P*, *p*, *P*, *p*....) of differential state of representing zero-status-stable mass-energy frame quantum particles.

Even though at the momentary level, the flowing 'energy-Times' of any mass-energy frame proprietorially differ from others until it reaches a **full-fledged** determinable termed functioning as quantum-particle or tiniest object (*something similar to the ideas of a linguistical expression through voiced and script, in the form of intertangling distinctive letters, words and framing them to make the reasonable or sensible sentences, where all the individual letters are sensibly distinct from the combined form words and then unitedly form meaningful* sentences.)

{Here, the termed **'full-fledged'** may mean something like the spectrum waves of the electromagnetic world, where one type of wave is not enough to distinguish the entire electromagnetic spectrum from the lowest to the highest frequency, generated from different equipment from supplying input *(something like **unfledged**)* to a reasonable output *(that is, **full-fledged**)* of different wavelengths.

Somewhat like the sense of ordinary use, a unit of 9-volt battery sourcing power supply, component levels' or integrally set elemental multiple ways functioning (*that is, **unfledged** works*) derivement, which act as the overall outcome applies some limited way desired (*that is, **full-fledged** product*) purposes.}

Accordingly, following the multi-dimensional roles of internal and external physics, the spectrum waves generated at the elementary level of energy elements, cannot accurately determine through the classically measuring methods, which generally calculate an overall outcome of the frequency spectrum as the waveform that gives only some justifiable low or high conceivable effects.

Like the Sun, a massive mass-energy frame integrated with uncountable energy elements with inconsistent frequency spectrums generating Gravity, and so the Gravitational field strength, besides several known and unknown releases, has comparably high or low potential than its other family members like planet Mercury, Venus, Earth, Mars, Jupiter, Saturn, Uranus, Neptune, Pluto, subfamilies, and also the distant planetary releases.

Also, it will be impossible forever unless the universal-Physics of the 2nd-energy-belt, natural ways cannot organize the diversified grades of energies in cloud forms or coagulate or desublimate some amount of energy of appropriate areas for further preparation of micro levels' particles, which consequentially become convenient to perceive through specifically designed bio-intellectual and many other non-bio living species of alien inhabiting atmosphere.

Likewise, an important point may quote following the remote technical sense of energy point that, to organize the diversified state of seemingly free-flowing energies released through the most-micro level of particles or their union as a large object or the small or massive dense energy-cloud forms and then to turn again into a state of coagulation or desublimation following the roles of muti-directionally deriving energy-Times, which play the most significant functions to act appropriately obeying the universal-laws, initiate with the growing process from birth and consistently amending (*if necessary*) even upon the maturity of the 2nd-energy-belt.

By the way, here may reemphasize from the technical viewpoint of energy, there is so much unimaginable diversity persists in the matured 2nd-energy-belt, even though the universal laws of the 2nd-energy-belt never let to break down or collapse since its amendable rules execute to retain balancing status while the 3rd-energy-belt and 1st-energy-belt never impel any external diversity for disturbance of the eternal proprietorially multifariousness roles in the 2nd-energy-belt, and also the unalterable in-born buoyant state of the Universe in the womb of the mother Supreme - Universe.

So, technically, the status of the matured Universe is eternal and unchangeable, and no matter whatsoever, its living and non-living creations display diversifying roles.

However, some ideological placement by the specific way set creatures, like human beings of this bio-Globe, calculate and decide based on some delimiting accessible coordinates and predict the destruction age of the Universe.

Indeed, from the standpoint of energy and following the law of energy conservation, there's the possibility of destruction of an observable Universe, so it does not mean the total collapse of the 2nd-energy-belt and its creations.)}]

ii. ►Gravitational field: -

In classical mechanism, a Gravitational field is a physical quantity recognized as the 'force field' that exists in the space surrounding a body or group of finite mass, in general, used to explain gravitational phenomena from all directions.

iii. ►The 'force field' of an object and a mass-energy frame: -

By the way, a little may quote about the sense of 'force field, whatever definition as the invisible barriers of force or coercion following the functions of observable objects.

At the energy level, this 'force field' may justify as some dignifiable created barrier or 'shield of energy' by energy-field due to consistent integrated elemental values or physics values roles that determine the positive and negative perspective of security, necessity, and likewise so many multi-directional purposes of the surroundings of the occupying areas, or field of a zero-status-stable mass-energy frame, regardless the mass-energy frame act as a tiny particle or massive object in a reasonable energy-Atmosphere.

iv. ►Elemental field on the point of elementary level of energy.

From the elementary level of energy elements of any referenceable (*that is, zero-status-stable*) mass-energy frame/s, the occupying inner areas by the energy elements can present as the field or the Elemental field. This 'Elemental field' may resemble the Gravitational field that proprietorially functions as an overall one type (*that is, gravity)* of force field.

This specific force field to an advanced stage's compact mass-energy frame/s can represent similar to the gravity-generating object/s, during which the overall Elemental values release functionally act with determinable pulling effects due to works of integral energy elements when the unit-value mass-energy frame/s (*or object/s*) perform as a zero-status-stable in the balance of internal (*constitutional*) and external (*surrounding*) physics values multi-dimensional applications.

By the way, one of the most significant points to note is that most of the observably fixed measurable values, which are commonly recognized through the word senses, may not truly work at the grass root level or elementary level of energy element of a mass-energy frame since the constituent motion element of any substantial mass-energy frame has a most fleeting role.

→ {Technically, from the viewpoint of energy, the fastest momentariness of integral motion elements in any zero-status-stable mass-energy frame or object tends to determine the position of zero-status-stableness in the surrounding standard (*parametric value*) atmosphere.}

So, it does not follow the standard parametrically measurable tools, which generally apply to observable items based on the delimitation of contemporaneous consciousness set in natural ways for the intelligent living's physical system.

v. ▶ Orbit, Orbital lines, and Orbital-Field areas:

In simple word terms, an Orbit is a path in which one object in space takes around another. The word 'orbit' quite often applies to space technology, somewhat like our solar system where the planets orbit around the sun, moons orbit around the planets, and similarly, objects follow the orbital path that takes around another.

Following a general conception, the aspect of an Orbit is justified as two-dimensional, and the Orbital is a three-dimensional ideology.

In modern physics, an Orbit is considered a fixed path along which electrons revolve around the nucleus of an atom when the orbital (*termed as wave function*) is an encircling hypothetic region of the atomic nucleus.

→ (*But, from the componential viewpoint of energy, an Orbit is far beyond classical definition, but possible to analog to the classical mechanism even though the appearance of the energy level's orbits is unlike the observable items that can comfortably assessed through the common delimiting conception.*

Therefore, Orbit, Orbital lines, and Orbital-Field areas at the elementary level of energy elements maybe remain imaginary to the modern bio-intelligent species, like the many physically unreachable hypothetical and mathematical complications.)

→By the way, it may requote that the works of diversified grades of energies in the 2nd-energy-belt, not delimited to some specific dimensions for the coziness to particular way designed bio-intelligent beings, directly capable of discerning unto some accessible extent which called as practical means.

From the dimensional perspective (*elaborated in part 2 of this book*), it may recall that any proprietorially distinctive energy element in a mass-energy frame may function as a minimum one-dimensional role following a 0-base *{(mean physically two-dimensional work, one acts as a base (0) and the other one for performance as mono state assignment (drawing EU39) with indication of some significant points to convince}* due to inherent or self-bearing Elemental values.

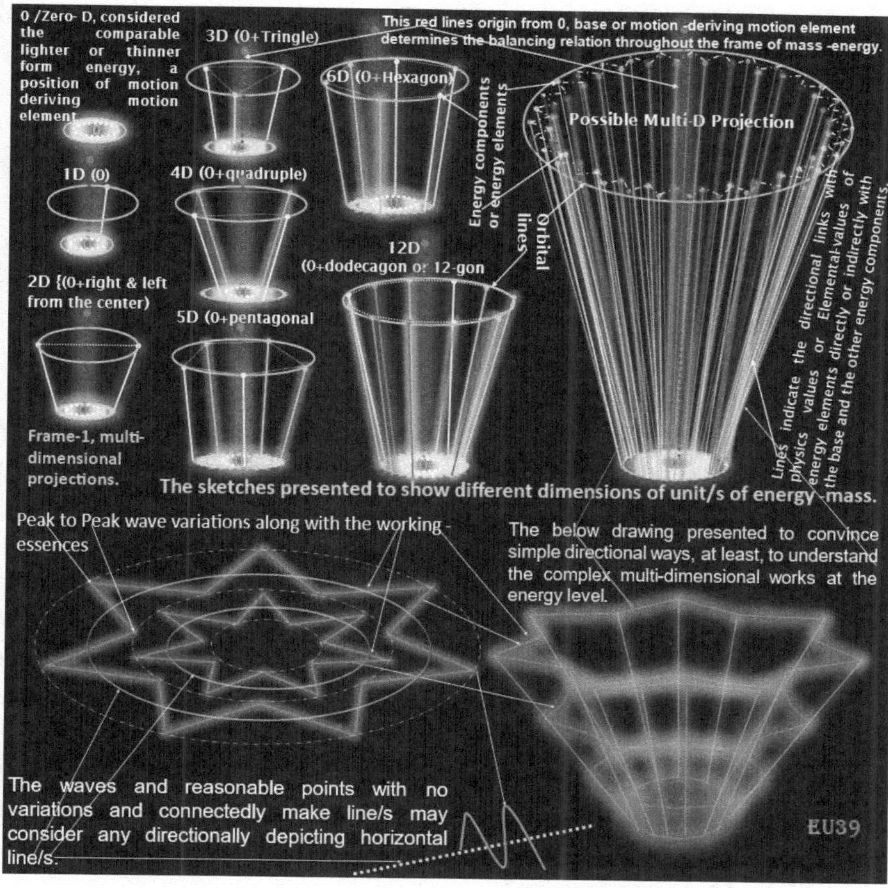

→So, say Elemental values, physics values, or elemental-Fields (‘*Elemental field’ also may present as the occupying functioning areas of Elemental value or working essence of an energy element*) of energy elements in any zero-status-stable unit-value mass-energy frame, either of fleeting appearance or long-existing feature as an object, capable of generating internal physics values in an organizing manner due to works of integrated energy elements in the zero-status-stable residential unit-value mass-energy frame, in coordination with the surrounding atmospheric-Physics and as well, for integrally or at component levels, the residential mass-energy frame or a prospective object as zero-status-stable unit-value mass-energy, which in a calculating manner can check, secure, and maintain a reasonable level of balance from the external application of surrounding physics.

(Remotely, the above energy levels exertion is somewhat similar to the works and exercise by the physical system that assesses through human beings' consciousness. Whereas the energy levels, many of the unitedly deriving body physics from the integrated healthy molecules of the weakly conscious human ill-physique, tend to manage attention up to some level of officially recognizing through the subconscious mind of the person. While many other living beings effortlessly or applying minimum effort can perceive the same thing when performing a healthy consciousness.)

Therefore, following the common physical sense, it will be logical to say that any unit-value mass-energy frame intelligently works due to the prompt and appropriate actions from their interlinking Elemental values (*Elemental-value is the True root cause of deriving multi-types Physics at the componential or elementary level of energy-element*) works of integrated energy elements that occupying specific functioning areas may call as elemental-Field/s of individual energy elements.

→With a drawing (EU40), it may try to convince that in a zero-status-stable unit value mass-energy frame, following their extent of self-bearing gravity, that is, 'energy-Field' (*in the drawing 'energy-Field' shown as peripheral areas*) of a mass-energy frame, its constituent energy elements make them organized on some imaginable circles that may call as orbits (*like the 1 to 9*) in coordination with the roles of motion elements and other working energy elements throughout, following their componential quality (*that is, Elemental values or Physics values property*) in any reasonable external circumstances, that is, applications of appropriate atmospheric physics.

So, any zero-status-stable unit-value mass-energy frame peripherally covers some working areas due to consistent internal physics values or Elemental values release, which may call 'energy-Field' act as the functional essence of the unit-value energy or proprietorially denotable identity of the individual object/s.

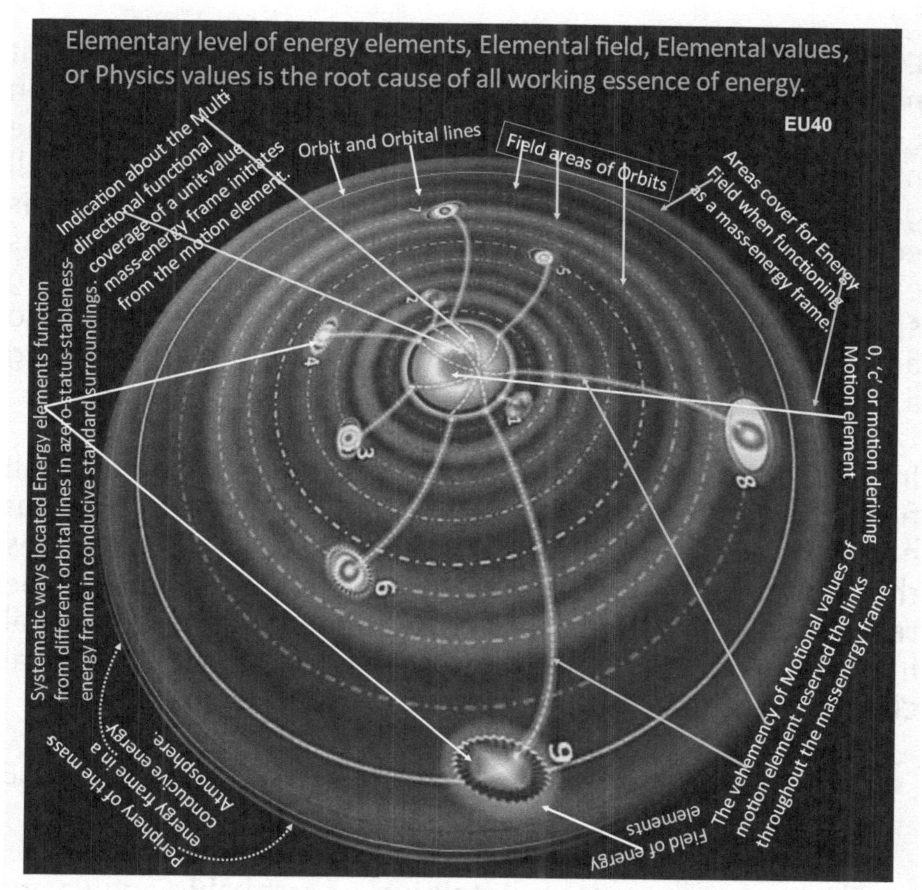

15.7) Gravity and its affiliated accessories' outstanding application in the association football prove the quality of intelligence of the human species concerning the techniques and better utilization of materialistically emanating gravity, besides some natural physics and non-natural ways deriving physics in this habitable Globe.

Further, for ease of understanding the application of Gravity, Gravitational field, force field, orbit, and orbitals at the elementary level works of energy elements to an overall mass-energy frame which represents an object, may try to convince analogously, applying some remote energy level's sense and following some duly exercises of the association football or soccer game.

Looking at the Three-dimensional perspective, indeed, it may find lots of similarities with the formulaic presentation E=mc^2 or augmented state as E=mc^<2> and the managerial arrangements of playing areas of the football ground, two teams or two ways of emotional creations concerning the diversity, gravity or attraction (or assume as emotionally deriving attraction force by the bio-living species) and interest among the players, spectators, socio and geopolitical dignity by inciting challenges for winning a game.

Indeed, from the 3-D outlook, the projection of the football field and the formulation of 'E = mc^2' are observable but not Truly similar in performance. Instead, it acknowledges the best gift from mother nature bio-Globe following the resonance of 3D perfect discernment and structural importance of human species evolved from the biological class of Energy in a bio-native inhabitable atmosphere.

►Following energy levels, it clarifies from the formulaic E =mc^2 as: -

→'E' acts as constant, maintaining unbreakable equivalent relation with mass (m) and speed (c) regardless of turbulences or diverseness within the specified areas indicated by mean of a power symbol and "2" or, more or less of 2, which is reasonable to analog with the projection of football field restricted within specific area matching like "E" the acting constant.

And, for football games, two teams represent two sections of mass (*as 'm'*) and derive momentum (*as 'c'*) for the liveliness of the games.

Rules and regulations provided by the management for defending and attacking, which act like some levels of 'c, speed or momentum deriving elements, encourage winning the game.

i. It concedes that the association football is one of the global sports in the world, gravitates to sociological importance since it interrelates with some form of pride and finance (*that solves some worldly desires*).

→In this day and age, based on the modern fantasy ideology, beyond the discernible sense and from the stance of energy, the association football includes involvers (*here, the association football and involvers, supposing concordantly formed clouds of mass-energy frames*), which someway weirdly became associated with the game.

→Accordingly, for a game, football players will act as essential working elements of a team or the representing mass-energy frame (h*ear, football players act as something similar to the spot-on working essence as energy elements in a mass-energy frame*), then the '**football**' will function as an unwanted element.

On the point of one's innerly deriving emotional gravity, for football players, '**football**' is like a loving heart. But following the principles of the game, there's no way the players will allow the ball to invade and cross the boundary of their team's goal or heart areas while the game is on.

Accordingly, players toil hard to emanate physics internally from their physical systems to chase and check the opponent's attacks in the **standard atmospheric physics** of the field areas and the football fans of the stadium.

*{(Here, the typical/**standard atmospheric physics** considered as the surrounding circumstances generated the encouraging (**positive**) and discouraging (**negative**) by the football fans. At the energy level, this directionally distinguishing property as '**positive**' may justify as the physics values' action and '**negative**' is the consequential effect or reaction that leverages due to implying some forms of action.*

(Nevertheless, technically, from the standpoint of energy, the objectively discernible idea of 'action and reaction' is some form of particular way derived directional forces, which is entirely unlike multi-dimensional roles of 'Energy' and its essential working components or energy elements.)}

[{Here, to make the principles of the game, administrative authority tactfully formulates the rules and regulations, which work as invisible tools /mechanisms for creating non-natural physics when the game is on.

Therefore, when the players are in offensive and defensive actions to abidance the rules of the game, they consistently generate non-natural physics from one's standardness of physical system *(here, standardness of physical system justified as the prescribed or natural capacity of deriving body physics of individual players)* in the conducive surroundings of the game.}

{Now, from the perspective of energy, in a compatible way, it can consider the football field as an acting mass-energy frame containing its limited working essence or energy elements like the 11 + 11 contending football players for performing physically entertaining delimiting directional sense as a 'positive' or one's favorite and 'negative' or challenging team.

Indeed, at energy levels' all functions happen multi-directionally, so physical purposes created rules and regulations by intelligent human beings generally follow some particular way or 3-D (*three-dimensional*) of natural ways developing perceptible capacity that satisfies the system upon getting an appropriate result after positive or action and negative or reaction activities.

However, the laws and roles of any form of mass-energy frame, containing types of energy elements following the universal law of the 2nd-energy-belt, intertangled directly or indirectly throughout, regardless of their individual quality of tangible or intangible physics, generated by any small or large unit value mass-energy frame.}]

ii. In addition, the training and development of the individual roles of the players in the team for maintaining intertangling or cohesive relations wherever their movements for the football in the field to make the team strong defender and champion against their opponents.

So, philosophically, from the physical perspective, players are entangled due to the pre-decided rules of a team they agree to play.

{Therefore, following the physical relations between the different positioning players in a football team may exemplify remotely the quantum-level **entanglement** practices of specific groups of microparticles or mass-energy frames they evolve following universal laws to solve distinctive purposes in the matured 2nd-energy-belt.

*(Concerning the sense of **entanglement**, the same is described in one of the earlier books based on linguistic alphabets).}*

(Further, the technique for organizing squads /teams and regulative bodies for the association football proves the quality utilization of improved brain processors set up in intelligent beings.

Based on that micro level's brilliance application of human beings, someone can easily understand the skillful handling of innumerable diversified forms of energies and energy products by the matured 2nd-energy-belt.)

→The boundary of the football field covered some rectangular areas of 90-120 meters long and 45-90 meters wide. Besides the rules and regulations of the game, many other associated items are pre-decided by some administrative bodies, like the International Football Association Board.

Whatsoever the above plans for exercising the football game, all programs and purposes, in a nutshell, may justify as conscious arrangements by some experts /intelligent planners for artful and entertaining games.

→Herewith is a drawing EU41 for refreshing the cognizance about the marking of an association of football grounds and the standard arrangement of players before the game starts.

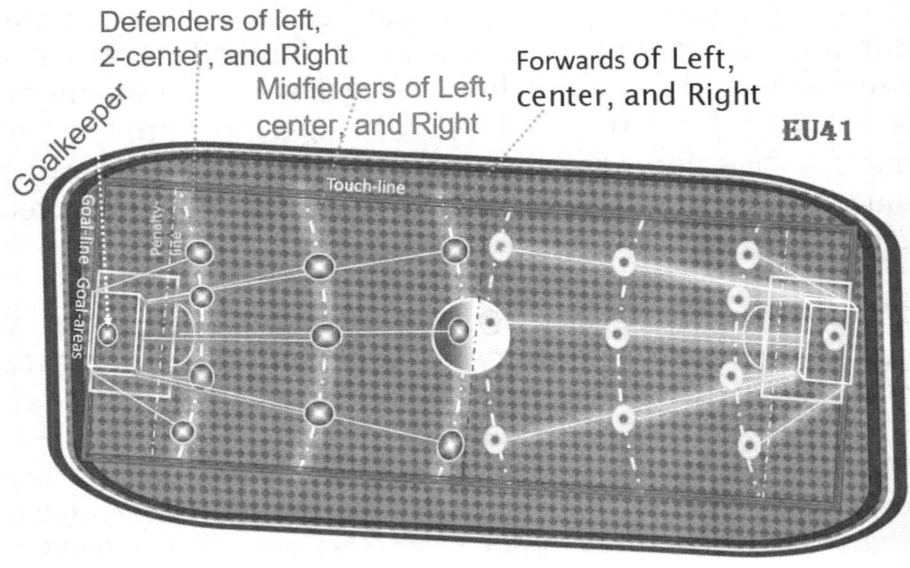

→So, for the human species, their particular ways of dimensionally discernible consciousness apply to many purposes, including enjoying different games. (F*ollowing*

the structural significance, this consciousness mainly rules through the 3-D physical projections as defined in earlier chapters.)

→But for the energy levels, it means that the physically appearing 3-D football ground (*or the football ground considered as the 'Energy-Field' for integrally or at component levels set energy elements*) is at the zero-status-stable positioning object or some state of a unit-value massive mass-energy frame when the standing areas (here, *standing areas means, mentally and physically trained influenced areas somewhat functioning like natural way developed Elemental values coverage of energy elements*) for Goalkeeper, Defender, Midfielder, and Forward for players those who represent the alikeness to proprietorially independent energy elements secured or covered by self-bearing working essences, that is Elemental values.

{(*One thing to remember from illustrations of earlier books is that energy elements have vast interchangeable properties due to consistent physics values exercises when a unit-value mass-energy frame or object is parametrically balanced and secure by zero-status-stableness in* ***conducive*** (**conducive** or parametric coverage) *physics values surroundings.)*}

iii. Now, it attempts to convince, analogously following zero-status-stable mass-energy frames and its constitutional elemental configurations, those somewhat presuming as similar to the entire arrangement of the football game with limited numbers of football players, associators, accessories, and all those who are directly or indirectly involved with the play, comparably follow like a zero-status-stable imperceptible cloud of mass-energy frame/s, representing the entire system of the association football.

→Besides the entire arrangement of the football game, team **administrators**' preliminary decides who will be in the forefront players and types of offensive and defensive actions while the game is on.

(Concerning the point of **administrators'** decisions, which *is somewhat remote like the universally applicable physics values roles that naturally organize some appropriate amount of energy for making a mass-energy frame, including internal placement of constituent energy elements based on their quality and befit to face the forefront functions upon typical external physics values application.)*

→Again, comparably presume that in the association football, representing diversified ranking individuals are readily perceivable objects or zero-status-stable mass-energy frame regardless of consciously accept with distinctive living dignitaries or other non-living identities *(or all of those, irrespective of living beings and non-living status representing as objects, matter, materials, etc., simple ways may consider them with the one natural title, that is 'energy products')* since all distinguishing and non-distinguishing items have direct or indirect connections with the invisible source, that is Energy or Energy-God.

→Also say, all compositionally set up with True sources of work-orienting energy levels' energy elements (*energy elements are the integral parts of even a most-micro mass-energy frame/s)* that may seemingly remote or implausible to many intelligent human beings, following some standardness of traditional mindset and some stages conceptually improved human species.

iv. The drawings EU42 and EU43 are the matching presentation of the EU41 to convince how association football games can interlink with the grassroots levels' works of energy elements of the mass-energy frame when designated by the personal rights while positioning them (*that is, mass-energy*) as zero-status-stableness.

(In also means that association football and all its related accessories are functionally independent in terms of their zero-status-stableness, much like small or large masses of energy in the states of distinct frames or clouds of mass-energy frames that have their limits in terms of personal rights to act in a self-dependent manner, following both the natural laws of physics directly and universal laws of physics indirectly.)

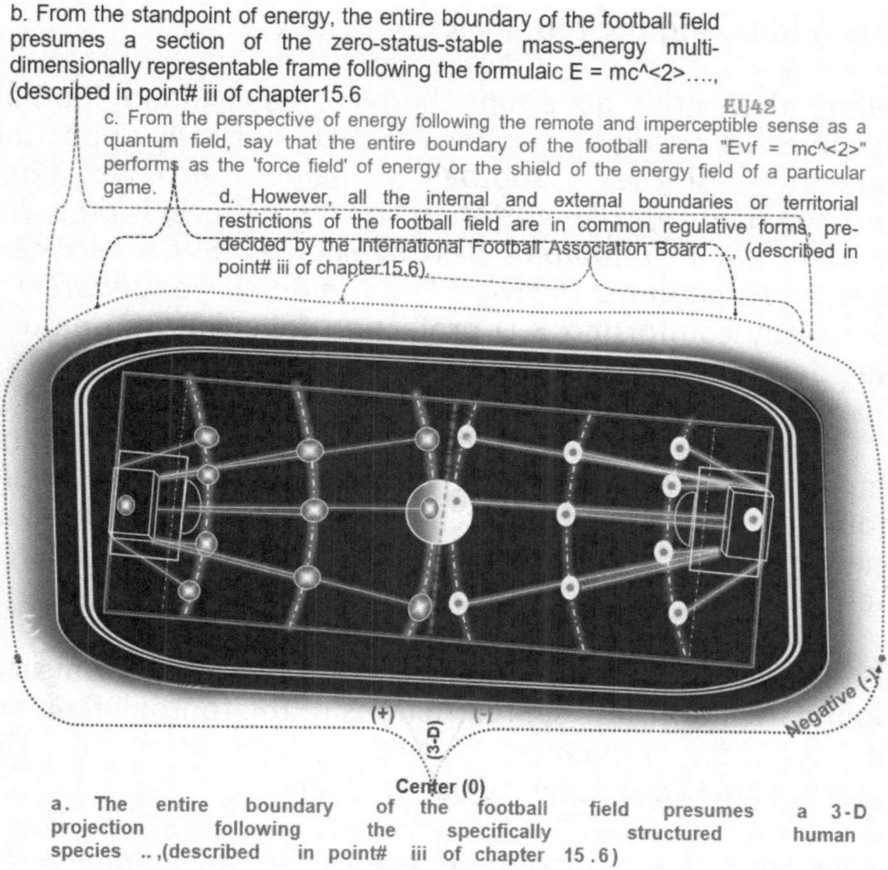

a. In the drawing and for application, the entire boundary of the football field presumes work as a 3-D projection following the specifically structured human species, justifying their commonly acceptable physical sense as right, left, and center, or the same may justify as the positive (+), negative (-), and center (0), that is three-dimensional (3-D) articulation for computational advantages for the intelligent physical system, regardless the factuality of multidimensional roles of energy components.

b. From the standpoint of energy, the entire boundary of the football field presumes a section of the zero-status-stable mass-energy multi-dimensionally representable frame following the formulaic E = mc^<2>.

*{*Here, the entire governing body of association football assumes as acting 'E' and for the 3-D physically appearing section of the football field mc^<2>, *(and the* segment *of the entire **arena** that covers only the football pitch or football field depicts as "Evf = mc^<2>" defined through the drawing EU23 in an earlier chapter)* a physically comforting 3-D projection following the natural way developed consciousness of human beings.*}*

c. From the perspective of energy following the remote and imperceptible sense as a quantum field, say that the entire boundary of the football pitch "Evf = mc^<2>" performs as the 'force field' of energy or the shielded areas of the energy field of a particular game.

d. However, all the internal and external physical boundaries or territorial restrictions of the football field are in common regulative forms, pre-decided by the International Football Association Board.

→But from the viewpoint of energy, it can somehow be analogous to the distant energy level's applications, where objects or zero-status-stable mass-energy frame's boundaries determine due to the application of universal physics values in concurrence with the surrounding atmospheric physics that acts following the constitutional quality of an object or mass-energy frame.

e. Further referring to drawing EU42 concerning the 3D indicated as a triangle, it may convey by adding a drawing EU42a that the brain processors of intelligent species are the exclusive gift from mother nature bio-Globe to abridge everything accessible in a direct manner and indirectly through some other means for three-dimensionally specialized (three-dimensionally or 3-D, basically act as self-bearing constant) human species........

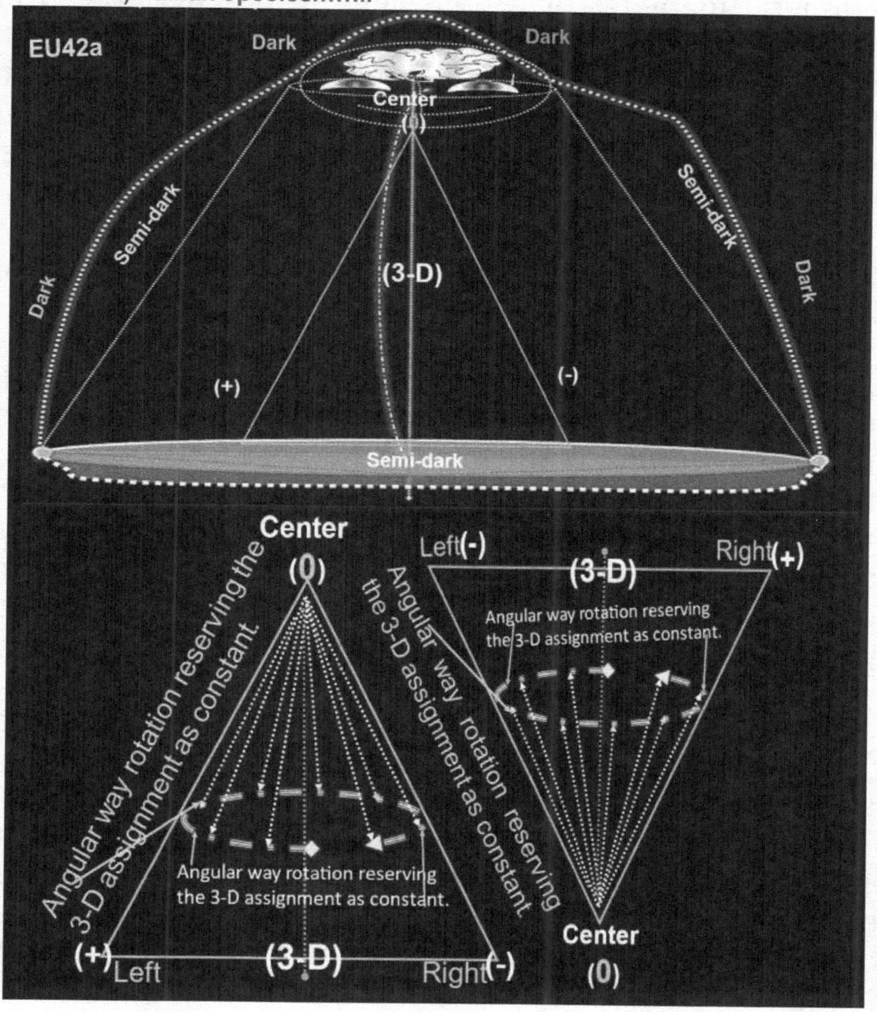

e. Further referring to the above drawing EU42 concerning the 3-D, indicates as a triangle, it may convey by adding a drawing EU42a that the brain processors of intelligent species are the exclusive gift from mother nature bio-Globe to abridge everything accessible in a direct manner and indirectly through some other means for three-dimensionally specialized (*three-dimensionally or 3-D, basically act as self-bearing constant*) human species.

And thereon try to abridge angular way rotating right-left, up-down, and also vice versa to the sensible limits of the peripheral diameter of a unit value living-Physique, from where the inducing multi-directional feeds give a conclusive outcome that satisfies individual, and possibly the majority of a group of a typically positioned species like human beings in these inherently caring conducive physics values applications from mother nature bio-Globe.

The same thing may happen to distant intelligent living either for an individual or group as a species.

→Also, beyond the coverage of 3-D access by the brain processor or something similar to that purpose of processing, for energy levels' many productive introductions, there's the tendency to reserve the right through the word-sense 'probability' or something like that concerning the 'indirect' consensus for semi-darks or dark zones functions since those areas behave as hurdles to living physiques for direct realizations.

v. In drawing EU43, the Physically representing Goalkeeper is assumed as the 'G' energy element or the acting motion element for the rest of the 10-member team or ten proprietorially distinctive energy elements, as and since the physically characterizing Goalkeeper (G) is the foremost responsible player for saving the Goal areas and goal, so the (G) became an acting Soul for the opponent's ultimate target.

→From the structural viewpoint for the external, that is, the peripheral shield of the Energy- field acting as a football field ($\boldsymbol{E}^{v}\boldsymbol{f} = mc\wedge<2>$), and its internally formulated borderlands like Run-off areas or $(mc\wedge<2>)_{Run\text{-}off}$, Goal areas or $(mc\wedge<2>)_{Goal\text{-}area}$, Penalty areas or $(mc\wedge<2>)_{Penalty\ area}$, Goals or (*mc^<2>*)$_{Goals}$ etc., work as different states of sub-shields of Energy- field for internal arrangements.

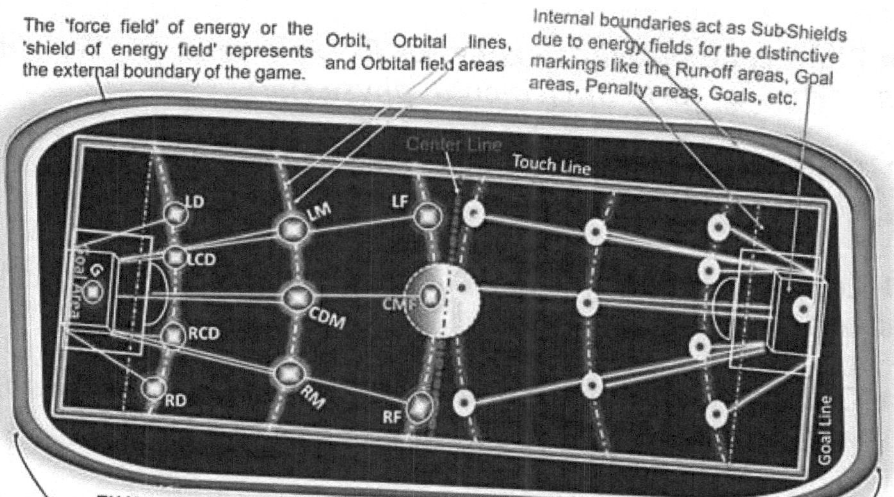

EU43

The peripheral shield of the Energy- field of a football arena (football arena assuming as 'Evf')

This same may resemble the natural way organized energy levels coverage area for functional demarcations of constituent energy elements (*or these energy elements at this moment physically assume as football players with distinct responsibilities concurring the differentia of the game*) of the **zero-status-stable** mass-energy zone (*that is the coverage areas of the football field as mc^<2>)* under a reasonable atmospheric-Physics.

(O*r, this* ***zero-status-stable zone*** is comparable *with actual playing zones regulated by the International Football Association board.*)

→While the game is on, the opponents have to overcome the chase of the defenders' ten other constituent proprietorially distinguishing energy elements or physically positioning players like: -

►Left defense zone *(or LD- element, for deriving specific elemental-physics),*

►Left central-defense zone *(or LCD-element, for deriving typical elemental physics)*,

►Right-central-defense zone *(or RCD-element, for deriving a particular form of Elemental physics)*,

►Right-defense zone (*or RD-elements, for deriving typical elemental-physics*),

►Left mid zone (*or LM-element, for deriving specific elemental-physics*),

►Central-defense-mid zone (*or CDM-element, for deriving specific elemental-physics*),

►Right mid zone (*or RM-element, for deriving specific elemental-physics*),

►Left forward zone (or *LF-element, for deriving specific elemental-physics*),

►Central mid-forward zone (or CMF-element, *for deriving typical elemental physics*) and the

►Right-forward zone (*or RF-element, for deriving specific elemental-physics*) to win by giving goal/s.

(*Or convey, at energy levels' constituent energy elements derived distinct elemental physics for some purposes to defend or attack for sustenance the set of mass-energy frame as zero-status-stableness.*)

→By the way, I think everyone will agree that the physical vigorousness of players is due to the **dynamic** roles of integrated molecules of the physiques.

Likewise, applying the remote sense may resemble **dynamical** activities derived due to the typicality of energy elements and their self-bearing Elemental values, which determine the functional properties of distinctive energy elements in the mass-energy frames, objects, or physiques like the football players who are selected for the standard position following their quality as defense, attacking or for both actions of the game.

vi. The drawing EU44 shows the standard standing position (*that is, before commencing the actual game*) of each proprietorially individual players like Goalkeeper, then the Right, Left, Center, and miscellany positions like the Defenders, Midfielders, and Forwards, etc., at the energy levels (*here, energy level means, while justifying the players as overall independent positioning zero-status-stable units of mass-energy frames with the distinctive roles following the given immediate responsibilities*) all those players assume as constitutionally functioning different states of working energy elements with self-bearing Elemental values that are ready to act as soon the '**imbalance**' occurrence from the standardness of external or internal energy-Physics.

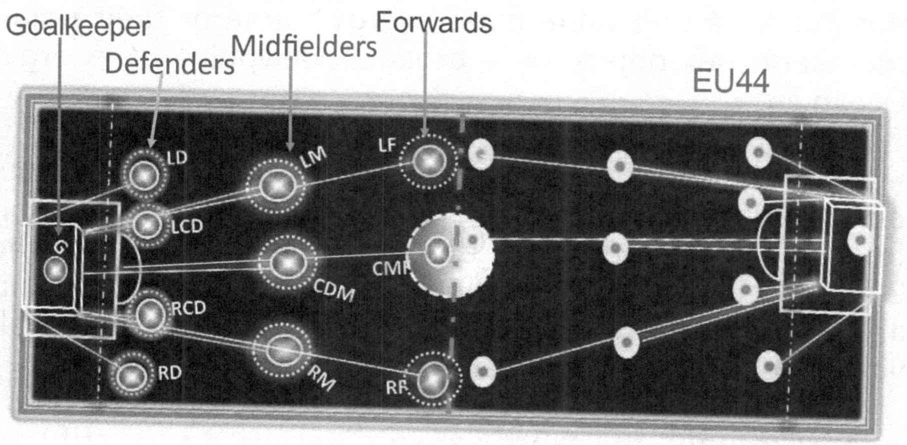

(*Here, for the football game, the word '**imbalance**' may mean the disturbance from some positioning standardness, initiates immediate commencement of movement of players for action, and the reaction of play following the administrative rules.*

But at the energy level, for a mass-energy frame, elemental values or elemental physics of constitutional set essential motion-element and other energy elements consistently in operation to held standardness of zero-status-stableness of the quantity of mass-energy areas it occupies in a conducive or standard atmosphere, and whenever get disturbed from the standardness due to some external or internal reasoning it gets ***imbalanced****.*)

vii. **At the energy level, Physics-values Apportionment.**

Also, it may requote a little about the 'energy-Physics' that justified the overall consequence of Elemental values works of integrated energy elements of a mass-energy frame, which virtually may be maintained as standard when the unit-value mass-energy frame qualifies to secure zero-status-stableness under the fair /reasonable external or atmospheric-Physics values apportionment.

Or, the same techniques may present as the apportionment of internal and external physics values to retain a balancing position of a unit-value mass-energy frame or their union represents an object in a broader viewpoint of energy-Ocean.

→Since the word sense **'standard'** usually (*that is, an observable mean*) applies as a mensurable value or position for comparable judgment, but for the energy level of the 2nd-energy-belt, momentariness does not allow to retain any such **standardness** status that recognizes as an overall virtual appearance for physical aspects.

→Accordingly, the word-sense diversity can justify as the eternal model for all aspects of the observable, non-observable, and any energy Universe.

a. Now, for any physically appearing expert player, their lively physique, irrespective of different operating organs and the regulating brain box, combinedly presume similar to a mass-energy frame. Accordingly, it can justify that this mass-energy frame became an expert player due to training through natural or non-natural means, either of direct invention from locale collections or indirectly through attainable **genetic** means.

Or, from the viewpoint of energy, this **genetical** inaccessible energy levels informative data that carry on from generation to generation upon the appearance of primordial building blocks for the preparation of most living organisms due to appropriate application of energy physics in a habitable atmosphere like bio-Globe with adequate life-orienting bio-grade energy resources.

(*Further, for conceptional clearance, technically, from the perspective of energy, energies never stay as still as the observable matter and materials, but the types of energy clouds occupy a small locale or massive cloud accommodated in an inhabitable nature like this bio-Globe is due to the contributions of visible and invisible residing and transiting multi-graded objects.*

Therefore, changes are evident whether modern human notions justify it ensue as beyond or within the parametrically distinguishing limits like extreme low or peak.

So, any dimensionally determinable status can never be factually permanent.

Indeed, the dignity of 'Energy' or 'E' is eternal, but their relatively related mass (m) and motion (c) always carry diversifying properties reserving equivalent relation with the fundamental 'E.)

b. Here, from the viewpoint of energy, nature's exclusiveness gift, like the regulating brain box of bio-intelligent beings, in common, function as living data storage, processing, and analyzing device for all the 'energy-Signals' derived as the form of information from the physique at the molecular level, while continuous dealings with the internal and external physics in a supportive atmosphere.

In a remote sense, it also may mean that due to the evolutionary process, the developed brain box, which became an essential regulative part for a physical section or the representing overall mass-energy frame for a lively human physique, integrally set with energy elements for accomplishing different works thus the players are capable of skillfully maneuvering the constituent organs and tissues while participating in the game.

c. {*Following physical practices, the sense of '**skillful**' is generally termed as the mental and physical fitness to perform sensible works.*

*However, from the standpoint of energy and different states of application of energy-physics, not only does any individual human being become **skilled** following the evolutionary process from prehistoric generations, but even its constituent molecules also get adapted and become proficient in tackling many unwanted energy-Physics, which determines the balancedness of physique* (or call the same as the healthy physique*) by apportioning the necessary internal and external energy-Physics, or the simple way same may define as sharing of loss and gain at the reasonable terms for the physical and atmospheric levels energies in facilitative circumstances.*}

→Indeed, for the players, the brain processor indirectly gets trained to regulate the body mass (bm) during play, which is somewhat remote at the energy level from comprehending in the observable manners that are relatively assimilable with the combining works of concerned physics values or Elemental values of different energy elements, and that altogether perform to maintain a zero-status-stableness of the unit-value physique in coordination with external atmospheric-Physics.

d. Whatsoever judgment makes following the accessible limits through the physical sense of intelligent human beings demanding dimensions as one, two, three, and so on.

But, following relatively analogous to the primary stage's some amount of 'Energy' that organizes gradually due to numerous applications of natural physics to make a frame of mass-energy zero-status-stable positioning physique, constitutionally set up with proprietorially or functionally independent energy elements, which unitedly tend to check and secure the stableness of the unit-value mass-energy frame/s from any multi-dimensional external application or atmospheric-physics.

→Also, from the perspective of energy, gradually toward the improvement from the baby stage's mass-energy frame, its constituent diverse energy elements gradually adapted for handling multidimensional maneuverability in coordination with atmospheric physics, which further efficiently allows to take advantage of supportive atmospheres and unite more and more appropriate mass energies, which eventually converted into most tiny stages structurally presentable object or particle/s.

And thereon, the advanced stage's formation of building blocks from different grades of life-orienting energy, besides lots of physics values application to give fundamentally significant living physique from a molecular level, then to their complex union with or without brain-box, somewhat quality wise justify as a no brain, good-for-nothing, useless, or especial with elevated degree intelligent similar to human being's brain-box that act as the bodily controller of integrated distinct types of organs.

→ *(From the perspective of energy about the earlier mentioned internal and external physics-value's ingenious application may distantly compare with the word-sense 'consciousness' that generally applies to human beings and many other livings when they act for something consciously due to some inconvincible reasoning.*

So, technically, energy or Energy-God is the most fundamental item for the derivement of any micro or massive outcome that appears through the physical substances.)

e. But, following the deeply rooted elementary levels works of energy elements in a unit-value mass-energy frame, regardless of most-micro status living or ungranted (*here 'ungranted' considered in general on the classical sense*) non-living sub-atomic particles, all releases are assuming some form of **energy-Signals** when they are consistently active to save and secure their zero-status-stable positioning under the conducive surroundings.

→From the perspective of the frequency spectrum for information, those released **'energy-Signals'** at the most-micro levels are completely special, and unlike the contemporary way defined through easily assessable, like stimuli that have meaning in some context for its recipient carries in the form of electromagnetic waves of frequency spectrums modulated message by varying the amplitude or frequency of the waves.

→So, on the point of the mass-energy frames, it may convey that a competent football player gets training deep down to the elementary level of constituent molecules, which conjoined due to consistent perfect communication while unitedly representing a physique. And accordingly, all energy elements of a total 'mass-energy' of any referenceable player become proficient in handling rigorous fluctuation of multiple energy-physics derived due to commending roles of brain-box and internal and external physics values applications while the game is in action.

viii. By the way, the predetermined arrangement of football players on individual teams (*assuming as Team 1*) may be analogous to the proprietorially distinctive elemental positions of mass-energy frames representing the players.

► Like the static territory (*assuming Goal areas*) for (G) or motion element has the acting physically responsible Goalkeeper (assuming Gk),

► and the interchangeable energy elements dignified as 'LD' areas (and *the physical aspect acting defender, the Left defense player or LDp*),

► the 'LCD' *areas* (*and the physical aspect acting defender, Left central defense player, or LCDp*),

► the 'RCD' *areas* (*and the physical aspect acting defender, the Right-central-defense player or RCDp*),

► the 'RD' *areas* (*and the physical aspect acting defender as the Right defense player or RDp*),

► the 'LM' *areas* (*and the physical aspect acting defender, Left mid player, or LMp*),

► the 'CDM' *areas* (*and the physical aspect acting defender, Central defense mid-player or CDMp*),

► the 'RM' *areas* (*and the physical aspect acting defender, the Right-mid player or RMp*),

► the 'LF' *areas* (*and the physical aspect acting defender Left forward player or LFp*),

► the 'CMF' *areas* (*and the physical aspect acting defender Central mid-forward player or CMFp),*

▶and the 'RF' *areas (and the physical aspect acting defender, the Right-forward player or RFp),*

as shown in the drawing EU45.

In addition to the above, the actual playing zone is segmented into two halves using a line mark (*center line*) for acclimating two contending teams, and arrangements of both team's players follow the pre-planned rules of the game from beginning to end of the game.

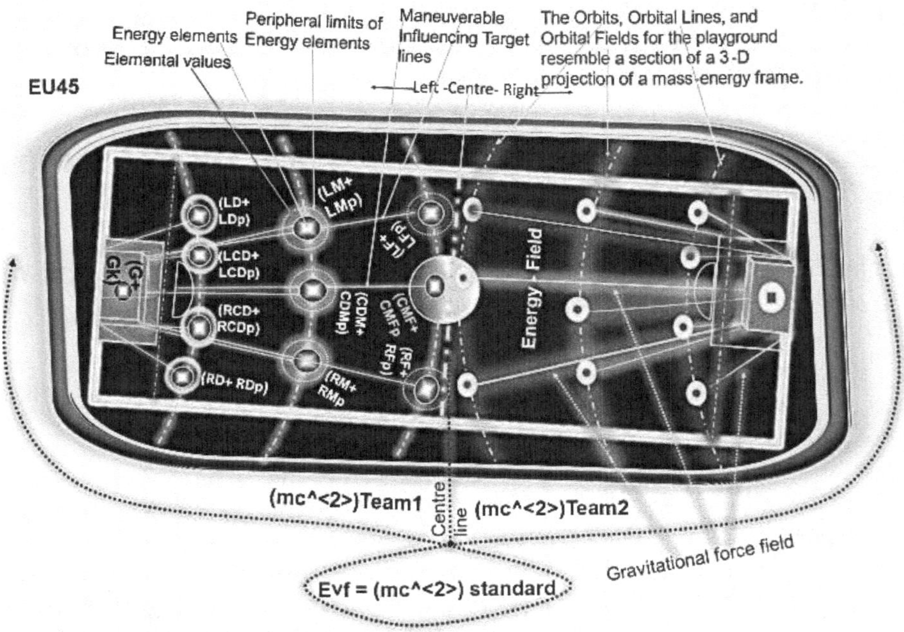

→From the viewpoint of energy and following the bi-directional mindset, the actual playground is justified as a standard unit of mass-energy frame, performing the Energy- field, ***Evf = (mc^<2>)*** *standard*.

And then, manipulatively for the two contending teams like Team 1 and Team 2, total 'Evf' share alternately two halves of duration during the match.

Accordingly,

► **Evf = (mc^<2>)$_{\text{standard}}$ can present as; -**

► **Evf = {(mc^<2>)$_{\text{Team1}}$ + (mc^<2>)$_{\text{Team2}}$}**

*{It also may mean '**Evf**' never loses equivalent proprietorial relation with (mc^<2>)$_{\text{standard}}$, its internally formulated* ***borderlands*** *like: -*

► Run-off areas or (mc^<2>)$_{\text{Run-off}}$,
► Goal areas or (mc^<2>)$_{\text{Goal-area}}$,
► Penalty areas or (mc^<2>)$_{\text{Penalty area}}$,
► Goals or (mc^<2>)$_{\text{Goals}}$, etc.,

work as different states of sub-shields of Energy- field for internal arrangements. And even whatever disorderliness happens between the contender (mc^<2>)$_{\text{Team1}}$, (mc^<2>)$_{\text{Team2}}$, and the individual players acting as functioning elements and elemental values during play.}

→ Further, for any team, assuming Team 1 will composite with players (*physical states*) or the representing energy elements like: -

► **(mc^<2>)** $_{\text{Team1}}$ = {(G+ Gk) + (LD+ LDp) + (LCD+ LCDp) + (RCD+ *RCDp) + (RD+ RDp) + (LM+ LMp) + (CDM+ CDMp) + (RM+ RMp) + (LF+ LFp) + (CMF+ CMFp) + (RF+ RFp)*

a. The drawing EU45 indicated with the lines of the Gravitational force field, which entice by the (G) {here, assumes that (G) works coverage areas that represent as *(mc^<2>)$_{Goal}$ areas* recognize as *Goal*, or at this moment presume overall Motion element "c" (*instead of mc^<2>*) that act as a pulling effect for rest of the **borderlands** for the Team 1 and even its integrated players those act as performing energy elements like *Goalkeeper (Gk), defenders LD+ LDp) + (LCD+ LCDp) + (RCD+ RCDp) + (RD+*

RDp) midfielders (LM+ LMp) + (CDM+ CDMp) + (RM+ RMp) and forwards (LF+ LFp) + (CMF+ CMFp) + (RF+ RFp) of a team.}

→Here, for the football game, it presumed that the distinctive energy elements act as the constituent players of a team, and the team justified as a union of workforces like the Elemental values of energy elements in a mass-energy frame or Energy- Field.

For the case of both energy elements and football players, the consequential act performs as defense of the team for saving the Goal or saving the significant position of motion element or 'G' in the boundary of the zero-status-stable unit-value mass-energy frame or durable object of energy-Field (*or playing field*) following the abidance of natural laws, which is somewhat resemblance with the non-natural regulating rules provided by the association football administrators for the players of different teams and the game.

b. From the standpoint of energy, saving the value of 'G' means preserving the Motional property of the motion element so that the zero-status-stableness of the simple or complex mass-energy frame, or the state of appearance as a matter with the distinctive property on the point of rigidity like solid, liquid, gaseous, plasma, and so on yet to acknowledge, could retain at the healthy position with proper control by regulating appropriate internal elemental-Physics or overall consider as energy-Physics, and by challenging the limits of externally applies atmospheric-Physics.

→It is somewhat similar to the cognition formulated for association football games while the game is on, all the players of a team tend to save the ball from entering their Goal areas or the heart zone so that the mental standardness of the players, administrators, and supporters of a football faction remain satisfied for no-loss and no-gain 'goal' status.

→Although the rules and regulations for the association football pre-decided, or considered a non-natural formulation by some expert authority. But, beyond the deep of non-perceivable and non-explicable sense, some unknown coercion entices everyone and assemblages as one partner from the involved players, team organizer, spectators, and supporters, regardless of the winning or one's favorite team.

(But, from the perspective of energy and quantum theology, I think some energy-level physics values activities call for the temporal union for a team of choice, regardless of whether someone justifies an indirect human being's creation as non-natural or direct natural energy level's works.)

c. By the way, some mixed form of analysis following the temporality of assessable objects, quantum particles, and unattainable energy levels physics values works.

In common, an object determines following some calculable dimensional roles.

→But for the multidirectional perspective, which mainly performs at the energy levels, and so for the cases of prevailing observable objects' physics values outcome.

Accordingly, for assessment of an object, in common, parametrically determine the balance, where internal energy compositions tend to maintain the consistent loss and gain practices following the diversifying roles of physics at energy levels, resulting in retainment of the question **action -and- a reaction** that utilizes as parts of determination or temporal computation about the assessable items through the consensus of intelligent beings by limiting the accessible areas and quality, when judgment makes for the broader understanding of works of energies in the fathomless energy-Ocean.

→For the 3D specialized human beings, objects follow three-dimensional outcomes whatsoever the constituent cluster union performs, following on an average positive and negative certainty to retain balance in surrounding support.

→Assessable objects are easily analyzable following some classical definition that satisfies the works of individual brain processor, either of one or their group consciousness.

→**But at energy levels, the boundary of any mass-energy frame determines its perfect Zero-status-stableness following the limits of internal and external multi-directionally arising physics in a place of locale or transit to different atmospheres.**

c1. Now, from the standpoint of energy, following the association football arrangement, overall areas of the actual playground can be considered as an energy field, mainly recognizing following the stance of 3D assessable objects, which fundamentally compute two ways of segregation like the alternately functioning positive and negative, maintaining a position similar to '0 /zero or something else as an unchanging separation line to cover 3D consciousness of an object or the playground.

→Next, the formulation of the (G+ Gk) of the football field somewhat can resemble the sense of core center, which functions as the origin or soul of a team and also becomes the target for an opponent. The (G+ Gk) also assumes an independent position with some quantities of the mass-energy frame, covering some areas considerable as a structurally representing object.

So, from the viewpoint of energy, this object and the areas covered by it are similar to the G+ Gk, considered an energy field, which acts as a union of quantum particles. For energy level, that particular section as 'G+ Gk' functionally presumes the motion element, which can emanate Motional value (*working essence*) whenever imbalance happens from a stipulated atmosphere.

c2. Consistent roles of motion element or (G) due to its working components motional-value or (Gk) resulting with gravity, and then demandable orbits and orbital lines in the Energy-field of the areas covered by the object having the property of zero-status-stableness allocated as playing zone for Team1, where based on pre-planned of the game, the players are provided with different roles for defensive or attacking actions of the game in their territory represents as the Team 1 or mc< 2> frame.

→Since motion emanated due to efficacious works of (G+ Gk), then consequentially formulation of the gravitational force field (EU46) determining the orbits, orbital status, and bonding relations with the other energy elements like physically appearing 4-Defenders, 3-mid-fields, and 3-forwards those stationed on their independent orbits, besides maintaining entangling relations wherever players move for winning terms to attack the Team2's territory or energy zone.

(*One thing to requote is that, since the physics value's 'gravity' is the constitutional part of a mass-energy frame, and accordingly, the resultant gravitational force or pulling coercions internally provided with some parametrically determinable bonding relations of multiperformance energy elements in a unit-value mass-energy frame or the representing object in standard atmospheric physics.)*

→In addition, one thing to clarify is that the sense of **'action and reaction'** is the befitting application for classically presented various statuses of matter like solid, liquid, gaseous, plasma, and microparticles or quantum particles that are accessible through some dimensional perspective and may satisfy intelligentsia of human beings and many other alien species.

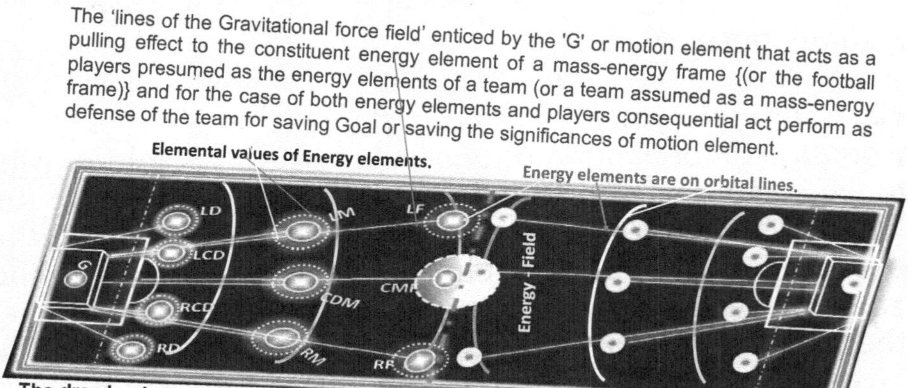

The drawing is a remote resemblance of the arrangement of skilled football players at different manoeuvrable locations in the playing field or presuming the energy elements set on the orbital positions in the energy-Field of a matured mass-energy frame or a representing durable object. Like, the static territory (assuming Goal areas) for 'G' or motion element is the acting physically responsible Goalkeeper, and the interchangeable energy elements dignified as 'LD' (the Left defense player), the 'LCD' (Left central defense player), the 'RCD' (Right-central-defense player), the 'RD' (Right defense player), the 'LM' (Left mid), the 'CDM' (Central defense mid player), the 'RM' (the Right-mid player), the 'LF' (Left forward player), the 'CMF' (Central mid-forward player), and the 'RF' (the Right-forward player). EU 46

But at the energy level, the sense of **action and reactions** may be replaced by the sense of loss and gain that maintains balance at energy levels in the 2nd-energy-belt. Whatsoever observable sense makes a sense like nothing and nothingness based on the limited dimensionally analyzing tools include core person/s to understand.

So, the word-sense reality that retains everlasting dilemmas following the observable items may give some form of satisfaction if judgment makes following the works of energies that perform throughout fathomless energy Ocean suiting as infiniteness, based on the capacity of the biological species.

→Also, everyone as a player or energy element, either influenced by natural means (*that is, atmospheric-Physics of a locale*) or non-naturally created rules (*by the administrative body of the football association*), has the target to save the individual and the team's territory by challenging the ball from entering the final target 'Goal, following **zig-zag** maneuvering techniques. The other way players move is to strike the ball towards the opponent's Goal area to win a game.

→ {(By the way, the sense of **zig-zag** maneuvering techniques is applicable based on some particular dimensional measuring methods, like naturally designed human beings are consciously capable of directly analyzing and determining while on three-dimensional activities.

However, from the viewpoint of energy, in the boundary (*that is, shielding barrier for the energy -Field*) of a zero-status-stable mass-energy frame or a durable object of a reasonable amount of the energy -Field, constituent energy elements tend to maneuver multi-directionally in an orderly manner in coordination with the multi-dimensional roles of atmospheric-Physics following the universal rules of the 2nd-energy-belt.)}

c3. For energy level, upon commencement of the game, the player's interchangeable movements (*partly shown in the drawing EU47*) from the static positions as defined earlier sketch EU44, like 'LD, LCD, RCD, RD, LM, CDM, RM, LF, CMF' all almost resemble where the energy elements in a complex mass energy frame, wherever their static positions on different orbital lines in some parametrically limiting standard atmospheric-Physics, intertangled through the manoeuvrable influencing orbital lines (*shown in the drawing EU47*) with the Goal areas or areas covered by the motion elements (*a most possibly thinnest state of energy or recognizing motion element along with operating Motional values*) in the mass-energy frame.

Somewhat like football Players, energy elements can maneuver from one orbital line to another following the versatile application of internal and external physics.

Presume as the networking arrangements in the football field or similar to the energy field in a mass-energy frame for safe maneuvering of players, or energy components or the energy elements.

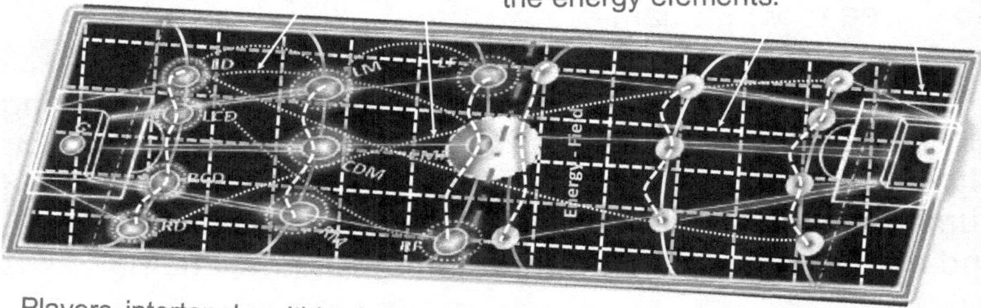

Players intertangle within the playing field following the created rules of the game. Hence they have been given the right to maneuver the football throughout the playing field while performing defense and attacking actions. Whereas the energy elements consistently can manipulate multi-directionally following the networking path since it systematized due to the balancedness of internal and external physics by keeping the position of the motion element (*the thinnest /lightest degree of energy in a unit-value mass-energy frame*) unchanged in the 'energy field' or comparable fine state neutrally flow energy in a compounded zero-status-stable mass-energy frame or a referenceable object. **EU47**

→In the case of a frame of the mass-energy unit, all those independent stationary positions of energy elements are naturally settled on the orbital lines when the unit-value mass-energy frame is justified as zero-status-sable energy -Field under the **favourable circumstance** (*or this **favourable circumstance** may mean about some extent of parametrically tolerable limits of external physics or atmospheric-Physics, but at energy levels, in the reality of Motional values' most micro momentariness, those logically applied parametric limits do not work flawlessly*), which is somewhat similar to the responsible players stationed for particular works before the commencement of the football game following the rules provided by the game regulators.

c4. Next is, immediately commencing upon the game, comparatively external physics get enhanced fluctuations from the earlier standard-ness (or the scenario stands as the enhancement of mental turmoil of players compared from, before the commencement of the game) due to the impressive support of spectators and many others who are directly or indirectly involved in the game and consequentially act in the physical system or consistent roles of internal corporal energy-Physics inside the players.

→The above corporal level's technical activities may somewhat resemble the integrated energy elements, those who initiate internal elemental values or elemental physics immediately upon fluctuation from a state of zero-status-stableness of a mass-energy frame in a standard atmosphere.

→Indeed, abiding rules of the football game are pre-decided by expert administrators, whereas, in the case of energy, everything this way or that way is interrelated throughout the 2nd-energy-belt, whatsoever positioned as a most-micro stage's mass-energy, the cloud of mass-energy, frame of mass-energy referenceable as smallest sub-atomic particles, big objects made up of recognizing matter, material or materialistic items, energy-Universe, observable, non-observable Universe/s, and many inaccessible things through the delimiting consciousness of bio-intelligent beings and their invented types of machinery to aid individual and group satisfactions for the specific way designed living systems.

ix. Some more clarifications.

(*The above corporal level's technical activities may somewhat resemble the integrated energy elements that initiate internal elemental values or elemental physics immediately upon fluctuation from a state of zero-status-stableness of a mass-energy frame in a standard atmosphere.*)

a. For the football ground, the assertive configuration to administer with Field markings like the Run-off areas, Goal areas, Penalty areas, Goals, etc., (*defined earlier*) is something similar to essentially confining accessories for the field of association football play, which acts as a standing placement that is, all fixed by non-natural rules determined by administrative bodies for the individual football players, teams, and as well the accessible logic applications of play that satisfy the essence of the game and entertain the fans and many other aspects.

→Whatsoever from the viewpoint of energy and analogizing the structural or organizational mindset of the football association, all of those Field-markings and areas of the football ground somewhat resemble simulation of natural-way shaped peripheral structures or frames and internal configuration for elementary level of energy elements of individual objects that proprietorially act with distinguishing characters when justifying them as overall dense state complex clusters of the mass-energy frames or diversified grades of packets of energy, externally secured by shielding and internally by types of sub-shielding barriers of energy-Field.

(*Drawing EU48 is an analogous presentation of the areas covered by the football Team 1 in the playing ground and a unit-value mass-energy frame ideologically considered almost matching framework, although geometrical perspective both are dissimilar.*)

a1. From the viewpoint of energy, it means that the 'Field-marking' areas function as territorial demarcation, something outlined by using objects, particles, or geometric lines (*that are the goal posts, nets, flags, chalk powders, etc.)* in the playing ground for the respective positioning of skilled /proficient individual players, those who are time-being the performers as trained individuals and may denote them as energy elements and the acting workforce for an overall compounded mass-energy frame of energy-Field (*that is, somewhat similar to the entire areas of the playground*).

a2. In a remote sense at the energy level, the field marks comparably may justify closeness to the standing orbital positions occupied by the characteristically performing distinctive energy elements, which **'arranged natural ways following physics values applications'** in a unit-value mass-energy frame when proprietorially assimilated for the overall functioning as zero-status-stableness of an object or a frame as a compact bulk-state mass-energy unit that indicated in the drawing EU48.

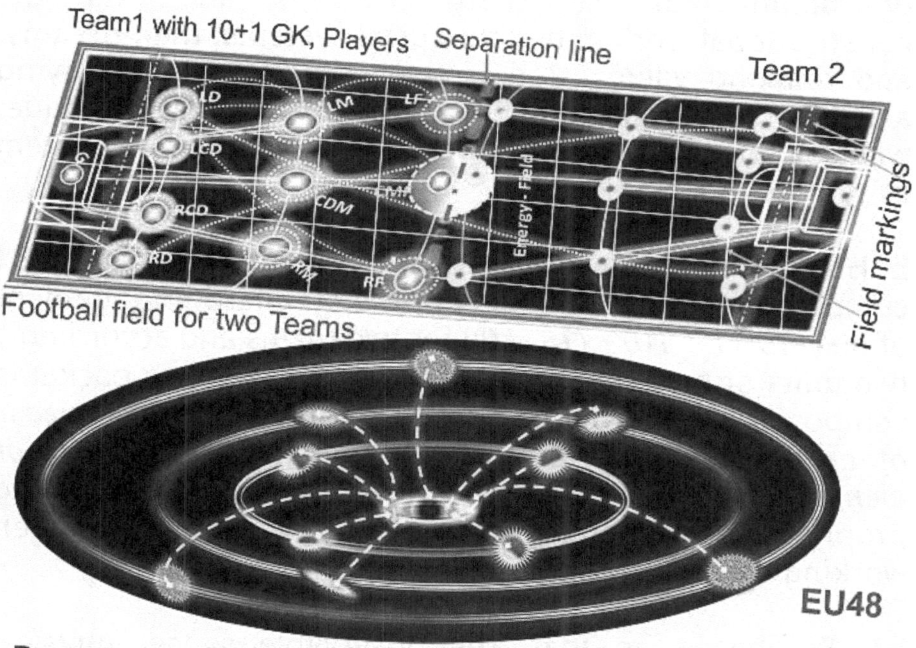

Presume a mass-energy frame is an acting Team-1 with 11 energy elements (centrally located motion element + 10 energy elements positioned on different orbits) like the 1 Goal Keeper and other (10) players stationed according to the planning of the game.

a3. [{(Here, for the football game, the sense of the **'arranged natural ways following physics values applications'** (*which is mainly applicable for the energy levels work*s) may remotely resemble the 'rules and regulations' formulated by the International Football Association Board for applications and to run the football game comes in an amendable package form for everyone directly (*players and many others)* or indirectly (*assume for spectators*) involved with the games somehow gets intertangles for compliance.

{By the way, recollecting the point about the word sense 'Physics' (*clarified in earlier books*) that traditionally applied to classical purposes may extend deep down the incessant elementary level's works of energy elements of energy, whenever and wherever representable with referential small or monumental mass-energy frame/s, and even the constitutional work in the biological systems, their dreams, and thinking values signals releases due to diversifying activities of internal building blocks of a living physique, whether analyzable or not directly through the own system or remote means.)}]

b. Herewith is a drawing EU49, where two sketches depict comparably a football field with the standard arrangement of players {(*1 + 10) + (1 + 10*)} for two teams and accordingly two units of mass-energy frames or objects, each packet is composed of 'GK' (*that is, Goal + Goalkeeper, or for the case of energy, motion element, and its working essence or elemental value or elemental physics*) + assuming proprietorially independent ten energy elements or ten working essences.

b1, So, based on that, the observable sense, physical construction, judgments, and many other things developed according to some inbuilt or preset measuring systems in concordance with energy physics of an inhabitable nature that mainly cover an overall 0D+2D projection of a living being.

Like the bodily recognizing function of gravity and many other word senses that deciphers similar meaning like attraction force and two levels of its directional roles. That means, while one functions as the positive of discernment, the other appears with a negative sway or anti-gravity to bring some level of balancing position satisfying an individual or group when applied for the formulation of many things, including the arrangements by implementing diversity with two teams, supporting and opposing spectators and likewise many associators to the game for association football, whatever the classically presented sense of 'gravity' is an attraction force between the objects since there are no functional prove about the practicality of anti-gravity.

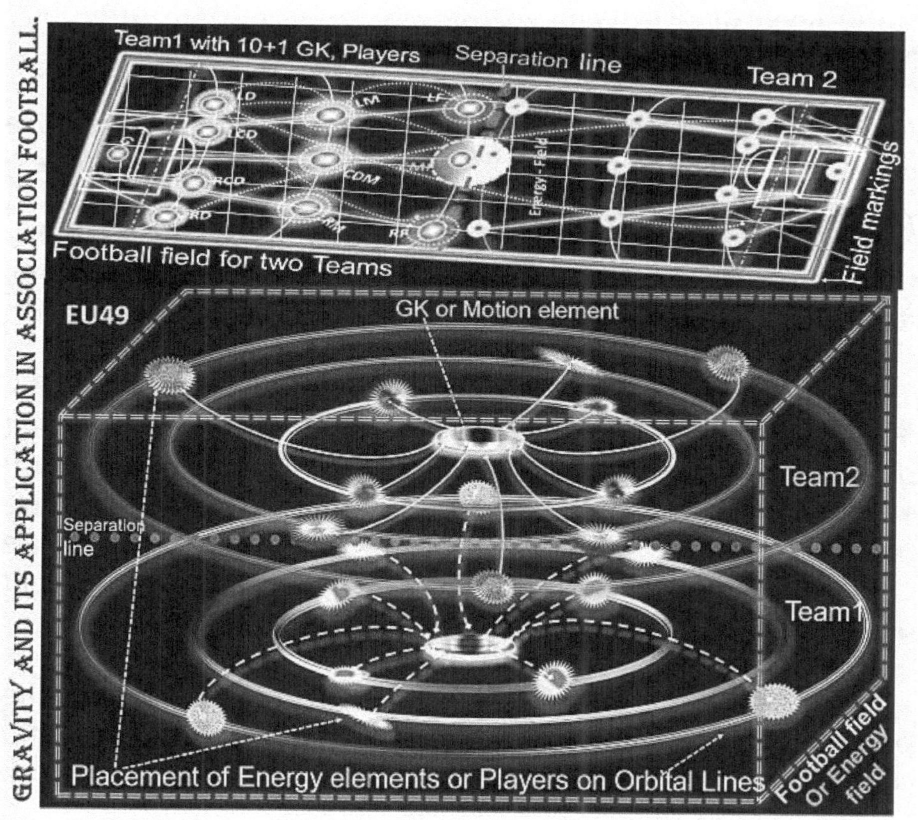

{Indeed, purely at the energy level, the 'gravity' is a constitutional part when some quantity of energy transforms (or assumes desublimation) *to mass-energy frame/s, whatsoever dimensional discernment follows particularity of structural values.}*

b2. Undoubtedly, the progress of the human physical sense is not a day's effort or miraculous appearance. But in lieu, the contributions carry forward from generation to generation in a place of survival like this bio-Globe, so physically adapted from the deep at the molecular levels of a union as a living physique in the conducive atmosphere/s.

The above point is quoted in short to mean and understand that any form of mass-energy frame, regardless of its micro to massive momentary existence, has the potency to yield some level of gravity or attraction force, either somebody judges on some attainable dimensions.

Accordingly, if someone thinks that human beings are nothing more than bulk states of mass-energy frames, undoubtedly, it is acceptable that the body mass of each human being is capable of generating gravity, except for the particular ways of application and recognitions following the structural significance and judgment through some dimensional analysis.

Besides, it is easily realizable that comparing the vast physics values activities in the energy -Ocean of the 2nd-energy-belt, how much unmeasurably insufficient structure of the human species in this bio-nature, like a molecule comparably expert than the multiple molecular unions that gave the figure of a bio-living physique.

x. The efforts of the total administrative body and the planners of the field entitled to formulate the rules and regulations of the game, the selection of players, and so on.

So, assume the administrative body and the planners here act as external performers /organizers since they are not directly involved in playing the game.

And for the spectators, those planners and administrative bodies are not functioning as players in the field while play is going on following all sets and subset's rules and regulations of the game (*or the same set and subset rules and regulations somehow may resemble energy levels physics that follows natural rules of locale formulated in coordination with the diversified activities of containing energies of the 2nd-energy-belt*).

a. Accordingly, somewhat from the perspective of energy, the same (*that is, combining external administrative involvers for the soccer game*) may resemble the roles of atmospheric-Physics (*i.e., Twig-physics + energy-Physics*), which is relevant to the natural way administering energy levels work for planning to organize the standard position of energy elements to perform in a simple unit-value mass-energy frame or complexly arranged with types of energies compact state unit-value mass-energy represent an object.

a1. So, an object, or considering an overall mass-energy frame, in general, or on 3D physical aspects of a human being, fetches two types of physics value's effects (*i.e., external natural implication and internal due to Elemental values consistent works*).

→Analytically, it also means if altogether the administrative organizations + players + playground + other accessories of football association +, etc., consider overall unit-values mass-energy frame 'mc^<2> function with the Superset and subsets, then its constituent domains of organizers + players + playground + other accessories, etc., etc., may consider as the distinctive working essences.

Like, **E= mc^<2> = mc^<0> + mc^<0.1> + mc^<0.2> + mc^<0.3> + mc^<0. 4>....etc^∞.,** (*drawing EU50)*

→Or may represent as: -

►Superset A0 (*or head/s of the organizer*)

= Subsets [{related to total involvers of Football association

= administrative organizations (A1) + players (A2) + playground (A3) + all other accessories (A4) +, etc.,}]

= an overall unit-value mass-energy frame (A) or an object, *or presume as the entire association football.*

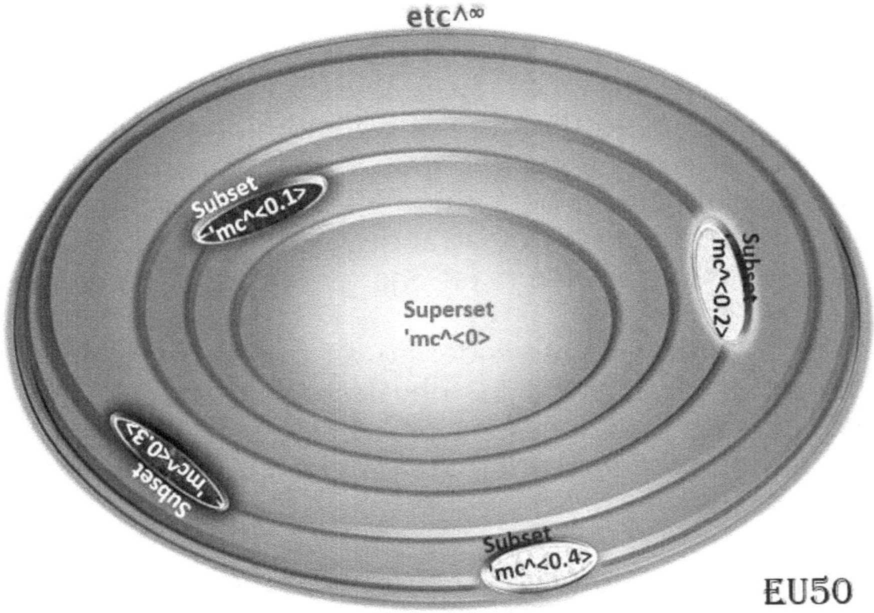

EU50

An overall unit-values mass-energy frame 'mc^<2> function with the Superset and subsets, then its constituent domains of organizers + players + playground + other accessories, etc., considered as the distinctive working essences.
Like, E= mc^<2> = mc^<0> + mc^<0.1> + mc^<0.2> + mc^<0.3> + mc^<0. 4>....etc^∞.

→Therefore, logically, when E = mc^<2> proprietorially works as a unit-value zero-status-stable position, its constituent superset and subsets as organizational A0 + A1+A2+A3+A4....likewise will be its functional or working essences.

Or from the viewpoint of energy, if the (A) is a compounded form of mass-energy frame mc^<2>, then its superset 'A0' and subsets, A1, A2, A3, A4.... and so on will act as individual domains of elemental positions like the proprietorially distinctive working energy element of the object, or complex compact state mass-energy represent an object (A).

xi. Further, a little more analytical elaboration about the configuration of association football games and the grass root levels' energy elements of simple or complex unit-value energy: -

a. The external boundary or peripheral boundary of the football ground is rectangular shaped and marked with two-sided long and short lines that restrict the movement of the players and the ball when the game follows the codified rules administered by the soccer administrators (*as shown before in different drawings*).

From the viewpoint of energy, at this moment, it may justify that the selected ground is an open and unoccupied piece of land or field (*here, the word 'field' commonly applies in the physical sense*) befit for playing football.

The land itself acts as some solid state of matter whatsoever the playing field experimentally defined as the soil is a state of three phases made up of solids, liquids, and gases, but on energy point soil is integrated with an unimaginable amount of compact diverse grades of mass-energy, which cover only a small stake of this Globe.

a1. Also, in general terms of classical Physics, the word 'field' represents a physical quantity that can be quantified following the provisos of scalar and vector values, mostly restricted to classical purposes for various states of matter.

But from the stance of quantum ideology, the word sense 'field' augmented to the energy levels that surpassed the feeling of vacuum, functionally True for conducting the activities of classical Physics.

Accordingly, from the viewpoint of energy, the arena (*the segment of the entire arena covers only the football pitch or football field depicted as "E∨f = mc^<2>" defined with drawing EU23 in an earlier chapter*) that covers association football may justify as a field of some state of mixed energies representing a zero-status-stable compact mass-energy frame, proprietorially integrated with energy levels works (*or means the persistent function of constituent energy elements*), or the same may regard as the consistent energy levels Physics-values balancing work, which allows of reserving the condition of soil suitable for the football games.

b. Concerning the word 'Gravity, it is delineated in many places in earlier books also, so I may give a little touch about the same for a better acquaintance of the topics and say, 'Gravity' recognizes as a physically invisible attraction force that pulls objects toward each other, and from the quantum viewpoint, it believes that the Graviton (*Graviton justified as a massless, electrically uncharged particle having the quality to act dual as a particle and as well wave*) is responsible for generating the pull or attraction force.

b1. But, from the viewpoint of energy, 'Gravity' constitutionally initiates at the elementary level of energy elements due to consistent active roles derived eternally by the Motional values or working essence of motion-element (*regardless it justify as the parametrically perceivable most micro level of potential or kinetical dynamism*) in a mass-energy frame, whenever the unit-value mass-energy frame represents a long or momentary zero-status-stable positioning, depending on the constitutionally occupying quality and quantity of energy, besides the stemming of internal physics values or Elemental values of the mass-energy frame in coordination with surrounding atmospheric physics (*or the natural way executing external physics in the ocean of diversified grades of energy*).

c. Now, considering the physical aspects of football organizations, including the piece of land, players for association football, and all others directly or indirectly related to the game.

c1. But, from the viewpoint of energy, it may remotely describe that constitutionally or indirectly by other means, all involvers capable of creating some low to high degree of effective 'Gravity' which may or may not be discernible or realizable as attracting force of the game, which makes the game tempting to involvers of the football association, players playing the game, as well spectators and other interested public.

→Since everything of physical appearance or referenceable as a mass-energy frame/s, built-in with some form of qualitative energy and at the elementary level of energy, consequentially capable of generating some degree of gravity or attracting cogency.

d. So, in the case of bio-intelligent beings, for football games, there are many criteria related to physically developed sensing systems, sociocultural obligations, surrounding circumstances, and many other things that may discourage them, and even not at all attract or can lure them by secondary means of the football game.

Accordingly, the sense of physical discouragement that somehow may assimilate at energy level roles of 'anti-gravity' or non-gravitational anti-pulling field means free from any gravitational pull and related effects, which commonly considered on the physical aspect.

Even though the word sense gravity, directly or indirectly, all are involved in creating gravity, whether it is a positive or negative application, and the same may apply concerning the football game to clarify for some people who are entirely or partly interested and not interested in games.

d1. But, for some uninterested people, due to leverage from the friend circles those who are interested in the game and regularly discuss, listen, and watch through media with all of their friends, the uninterested friend slowly gets inspired, gravitates, and unwittingly becomes entangled with the enjoyment of football association.

{(*Here, the conception of susceptible /sensible gravity mainly relies on the features of matter products that later developed due to the coagulation of classes of diverse forms of energy.*

According to the specificity of bio-intellect design, which mainly a 3D cogency, gravity on the sense of attraction or attractive-coercion, there should be some form of anti-value or deterrent coercion either to a status of a low or high degree of negative value, or the same on the physical aspect, say as the Anti-Gravity-coercion or repulsion whenever some internally generated yield {here, 'yield' may consider as some form of *thinking values' energy signals* devise after analyzing through a brain processor}, *act suppositive as positive of role.*

So, 'anti-gravity' is about the object/s being free from gravitative - coercion. But exceptions are the judgment through the brain processor, which overall delimited for 3D analysis and outcome, mainly exercises the negative, or anti, or opposite role of attraction, which at some stage turns into cynicism due to materialistic involvements.)

Whatever the different opinions about the act and effect of anti-gravity on physical and emotional aspects of materialistically influenced intelligent species. The holy soul 'ostensive Purest-energies' in the 2nd-energy-belt, entirely neutral and free from any virtually compelling acts due to gravity and anti-gravity coercion.}

d2. Analogously, the playground of association football boundary presupposes the peripheral frontier of an overall unit-value mass-energy frame representing an object of some conceivable existing state of matter.

In a referenceable unit-value mass-energy frame, energy elements are capable of functioning appropriately and maintaining the zero-status-stableness of an object or mass-energy frame following the formulaic (mc^<2>) as long as component levels' (*internally*) deriving physics (*that is, Elemental values or physics values*) reasonably balanced with implemented rules of external physics.

(*Indirectly, at the energy level, the word-sense rule also may apply for the imperceivable elemental values, energy values, or physics values since this word rule merely sets and adjusts as the needs arise.*

Also, the sense 'rule' indicates more flexibility for any application besides objectively implying from the low to the high sequels since roles of energies in the 2nd-energy-belt are eternally divergent.

So, the allowance of the word sense rules and regulations mainly allows for grasping some level of consolation concerning the direct physical roles of assessable items.)

15.8) When it comes to the concern of energy, the concept of Gravity and its natural implications on living biological systems in an inhabited place such as our bio-Globe is worth considering.

i. From the classical viewpoint, I think most modern intelligent living beings are aware of some fundamental points about Gravity that are: -

In physics, Gravity is recognized as a force that describes how objects interact by having some form of mass or (m) (*here, the (m) means occupying some space on dimensional aspects)*. The greater the (m) or mass, the greater the gravitational pull or attraction force from its counterfactor (m). Besides, the 'force' of Gravity decreases upon increasing their distance.

Also, Gravitational effects happen because of the curvature of spacetime and the movement of objects through that curved spacetime (*so, plainly can say, gravitational effects occur due to the diversifying role of motion of objects and spacetime*). And just like any other matter, gravity impacts particles like atoms and gas molecules.

(It's an intriguing philosophical perspective for someone who refuses to acknowledge that 'Energy' is the core factor behind everything. In such a case, the mental satisfaction of an individual would drive them towards the fleeting pleasures of materialistic or temporal life, where they can experience the illusion of being alive and active and can find an easy-going existence without searching for any purpose of life and one's true origin as energy or Energy-God.)

ii. From the above and the perspective of energy, it logically can depict that Gravity affects anything that accepts as an origin from energy and structurally performs either immature mass-energy frames or their matured state objects.

Even the very early stage's mass-energy frames are evolving to get a dimensional status, regardless of whether the tiniest or massive quantity of energy is dimensionally discernible as living or non-living rankings.

iii. Ensuing the above here may recall some technical extract from earlier definitions to convince the works of Gravity and its close application to living like the human species.

Indeed, following the observable sense, judgments based on some particular dimensional aspects (*shown in the drawing EU36 and earlier book in part 3*) like intelligent beings and their invented computational machinery, the word sense 'perfectness' and its true essence is almost impossible to imply since the observable sense always persistent with some form of stopgaps for findings the real solution on the aspects of Gravity and many other energy levels functions.

(***Since following the observable sense, objects work based on their delimiting dimensionally discernible structural values, whereas 'energy' tends to abide by the universal multi-dimensional laws of the 2nd-energy-belt, unless the natural physics values influence desublimate some amount of energy into a frame/s of mass-energy to act as a discernible independent position like a microparticle and then their union as a structurally configured big object like matter, material, materialistic items etc.***)

iv. Cleanly about energy, Gravity flawlessly works in multi-directional ways when considering the diversifying works in the fathomless energy ocean of the 2nd-energy-belt.

→Multi-dimensional roles truly function in balancing order with real no-loss and no-gain conditions at energy levels, which is unlike any dimensionally representable structural value like the matured mass-energy frames, objects, observable items, and even the living species formulated based on the conducive energy levels' physics, consistent with some form of loss or gain upon crossing the parametric ways of comprehensible limits that are restricted to stay or travel by the constitutionally deriving physics values cohesiveness. (*It means a reasonable energy level's areas covered by the 'mass-energy' as a frame to play with consistent physics values activities, having the quality to maintain as a whole unit.*)

v. Human beings are one of the best creations of this mother nature bio-Globe, formulated and advancing 'eras after eras' from the grassroots of micro molecules, then progressively became ancestors' living species following specific criteria that delimit in many aspects since they became specialized for three-dimensional discernible acumens following survival necessity to maintain and improve the cohesiveness of the structural values in the ever-versatile diversified energies all around.

vi. Following the formulaic ways presented, E = mc^2, where 'c' performs the Motional role whenever it represents the life of an object or respectful living being, became the root cause of all discernible or non-discernible energy levels functions.

Therefore, Gravity is also one of the outcomes derived when the types of energy elements in a mass-energy frame consistently perform in a conducive surrounding atmospheric physics wherever staying or traveling conveying as a frame of matured mass-energy, physically acting object or a life form living.

a. Conveniently use the celestial word-sense 'emotion' instead of classically depicted 'motion' that relates to the movements of various states of matter.

I think the word sense 'emotion' is the best option to distinguish and convince earthly or temporally deriving 'motion' from bodily emerging bio-class mass-energy frame representing a human being, entertaining living life while existing in this biologically comfortable habitable nature.

→ {Note, since the word 'emotion' takes the innerly spiritual position of the humanly intelligent beings' brain processor. *(Here, the word spiritual has nothing to do with religious sentiment. Instead, it is suitable to implement for energy-level applications.)* Thus, as an alternative, I find the conventional word sense 'emotion' is the best choice to distinguish the identical word sense 'motion, which is applied mainly for the other energy levels' Motional values functioning for all measurable or unmeasurable activities.}

a1. {While human beings perform any judgment through their constructionally entangling 3D fundamental factors, they naturally or automatically section the entire body mass into (1D) + 0D + (1D) = 3D (*indicated earlier with a drawing EU36a*), or the same physically may be defined as the right or positive (+) and left or negative (-) socio-culturally distinguishing synonymic and antonymic senses (*those are like good-bad, friend-foe, love-hate, etc.*), maintaining the center (*0D or 0 or parametrically measuring balancing point for (+) and (-) of discrimination*) as a neutral position for all analysis and computational purpose in the brain processor.}

a2. So, someone at least can guess that the emotionally deriving Gravity is due to the overall body mass of a human being, which delivers to the brain processor to regulate the system following the materialistic, worldly, temporal desire, or for some other reasons (*like remote influencing signals*) that mainly acts following the guidance of bodily significant sensing organs.

Therefore, in some cases, socio-culturally defined synonymic sense performs contrary to common understandings when the judgment endorses some physically influential sensing organs.

►From the above summarizing ways convey, it is reasonable to say 'Gravity' works deep in the human bodily construction that works as a decision granted by the brain processor and even affects all diversifying ways of entertaining purposes involving socio-culturally exalted or denounced. And, even that applicable for the association football from the level of organizers, teams, players, spectators, and all other involvers knowingly or unknowingly like as a tournament.

vii. Further, the concept of Gravity is directly or indirectly involved in all the crucial physics values roles, which can be tangible or intangible through observable objects behaving in the three-dimensional perceptible world following the 3-dimensionally evolved inhabited intelligent beings on this bio-planet.

So, 3D became the sole determining factor in how to justify the multidirectional movement of vast classes of energies in the 'energy- Ocean' of the 2nd-energy-belt, typically analyzed by biologically positioned living beings on this inhabitable planet.

a. For object and observable concern, every dimension (*dimension mean, an assessable scope for a particular kind like length, breadth, depth or height, etc.*) typically determines the distinctiveness for recognition through the 3D body-mass physical attainment of a living being. Also, some other additionally discernible degrees to the assessable limits apply through supplementary analyzation practices.

b. But when depiction comes to multidimensional discernment following the observable sense of object and various states of matter. It may find some acute inaccessible positions since, at that status, all possible discernible degrees either transform to some other states or dissolve into energy and its fastest versatile roles, resulting in fundamental levels' properties of mass-energy frames as objects collapse from original dimension, and may give result to some computational essence like the 0 findings, or 0abs^n or absolute nothingness as a sense of satisfaction.

c. Indeed, at energy levels, multidirectional activity performs with eternity. Since ideologically, it plays with no particularity. It means if using the observable scenarios to analyze some act based on particular dimensionally assessable degrees, the rest of the inaccessible probable dimension/s will cover the balancing works for energy.

d. So, at energy levels upon multidimensional finality, all perceptible dimension dissolves with the reality of energy levels fastest (*truly at energy levels*) multidimensional performance, where all objects, observable and logically discernible dimensions decayed, and the journey begin with new-new appearances.

→On the other way, multidimensional physics values application to some amount of mass-energy, abet to coagulate or desublimate to some state of object/s.

e. Living molecules also participate in some typical dimensional activities based on the quality of constitutional flexibility, unlike the performance of a living being's whole-body mass represented as a physique, which sensibly or regulative manner performs a Three-dimensional discernible lively mechanism while existing in a healthy condition.

►The drawing EU50a indicates the following: -

i. A complex scenario of a living molecule (shown in the drawing EU35) compositae with bio-class energy (*blocked in a cubical shape*) consistently deriving energy Physics while micro levels of waves or ripples incessantly interact to check and determine the zero-status-stableness as a state of appearance.

ii. Representing Soul or an acting constant, a comparable the thinner form of energy, where waves or ripples of energy perform their roles within the periphery of the mass-energy molecule.

{(The above point seems complicated when judgments perform following the three-dimensional pre-set computational essence to intelligent bio-living species. But I think it is standard to natural physics that reasonably regulates for locale or specific energy zone (like inhabitable

energy areas) following the Universal physics of energy - Ocean in the 2nd-energy-belt.)}

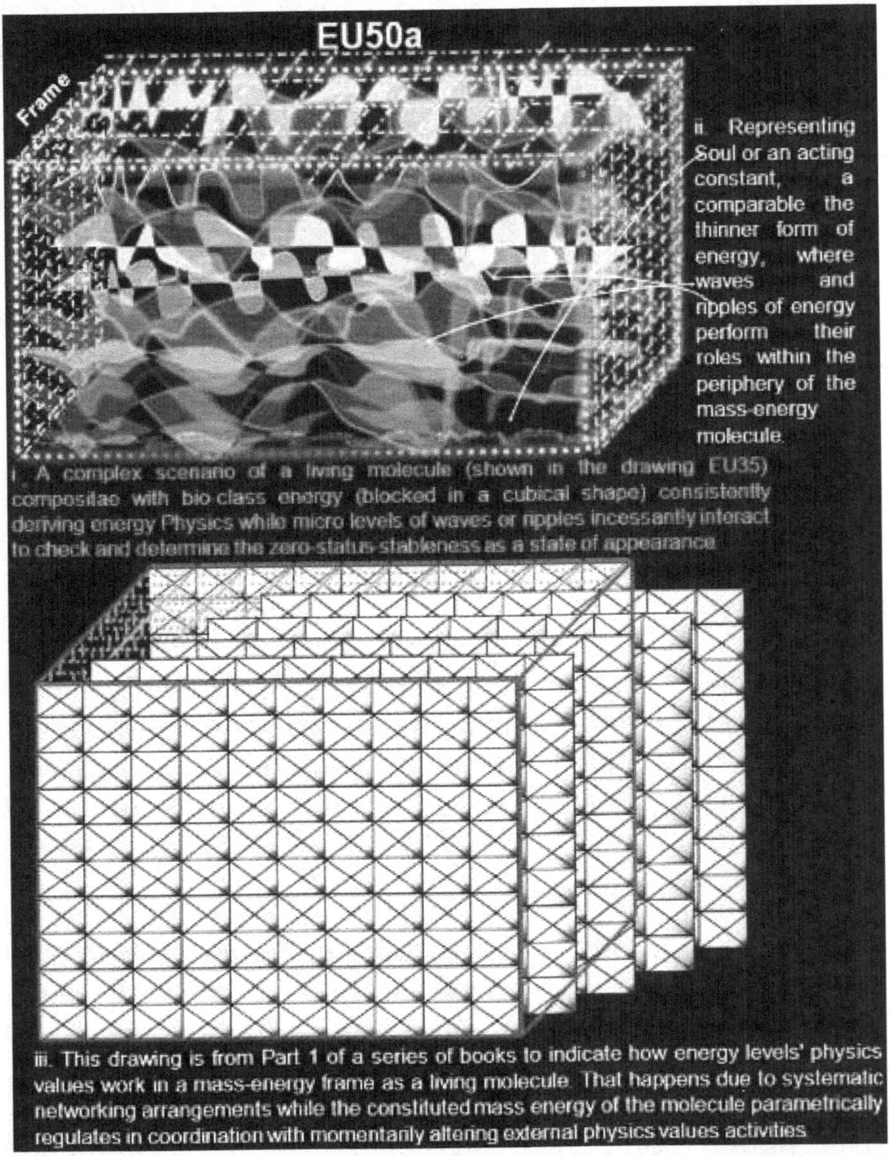

iii. This drawing is from Part 1 of a series of books to indicate how energy levels' physics values work in a mass-energy frame as a living molecule. That happens due to systematic networking arrangements while the constituted mass energy of the molecule parametrically regulates in coordination with momentarily altering external physics values activities.

f. So, beyond the 3D assessments through the brain processor of a living being, all other appraisal and discernments for multidimensional, in turn, have chances to narrow down or exercise as descending from the positive of discernments towards negative, from some lower degree of darkness to complete dense dark zones for 'analysis and findings' nothing (*or the same may mean annihilation from a state of acknowledging observable appearance following some dimensional assessments that is 3D + if any abetted dimension*).

15.9) A comprehensive portrayal of the interlinking relationship between the various aspects of physics, derived due to diversifying activities, leaps up and down at the different energy, quantum, and classical levels of physics.

In addition to the earlier definitions, comprehensively with a few drawings EU50b and EU50c, I intend to give a summarizing but some feasible touch about the eternal interlinking relationship between energy, quantum, and classical physics in the overtly diversifying activities of everything in the ever-juvenile 2nd-energy-belt.

► The drawing EU50b is a complex three-dimensional projection, exhibiting an overall coverage of the mass-energy frame, the acting body mass of a living physique, or an object that can be justified as some mixed state of matter, indicated with the following points: -

1. This drawing EU50b indicates how the physical system of a living being comprises different organs and molecules that overall represent body mass and at energy levels settled to perform some reasonable state of multi-directionally interlinking with the internal (*physical and*

energy levels) physics of body mass and external atmospheric physics (*Twig +Energy atmosphere*) at energy level, maintaining reasonable balance without losing a most-micro quantity of energy. But a lively physique performs Three-dimensionally from childhood till exhausted from liveliness as an appearance.

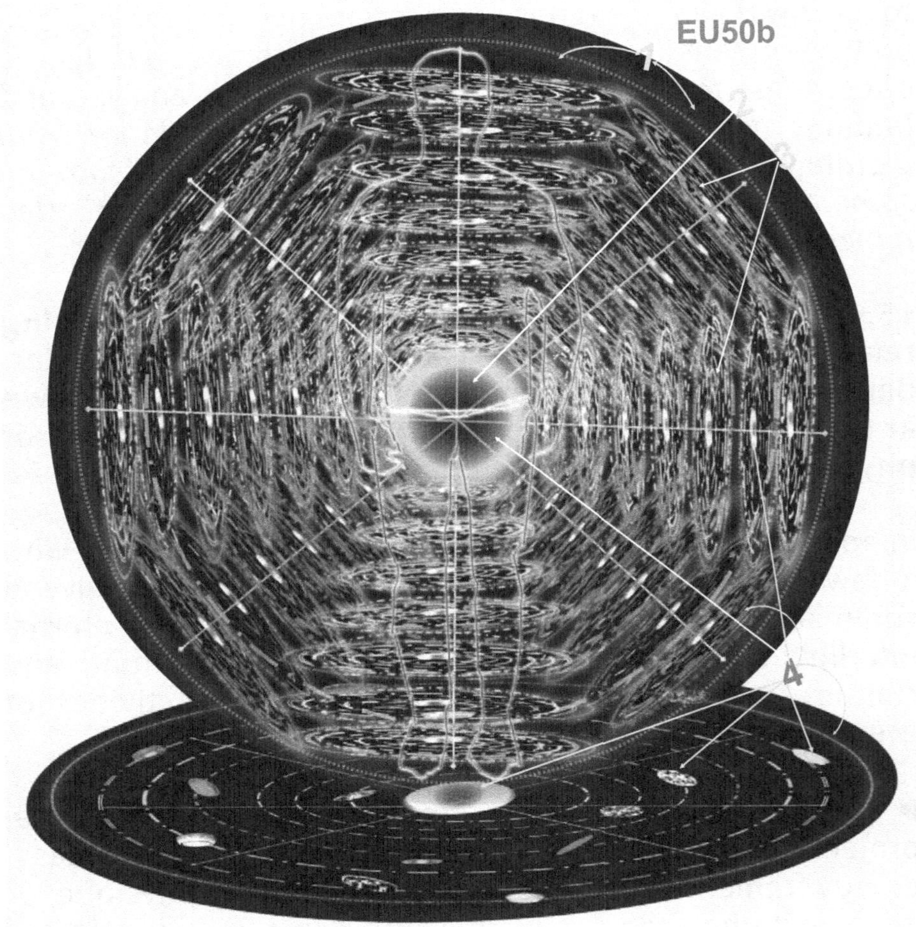

2. The Centrally aligned soul for a living body mass or a physique represents the energy as a complex union of multiple bio-classes of the life-orienting bio-grade mass-energy frame, following the formulaic representation as mass (m) and Motional values (c) interlinking tie.

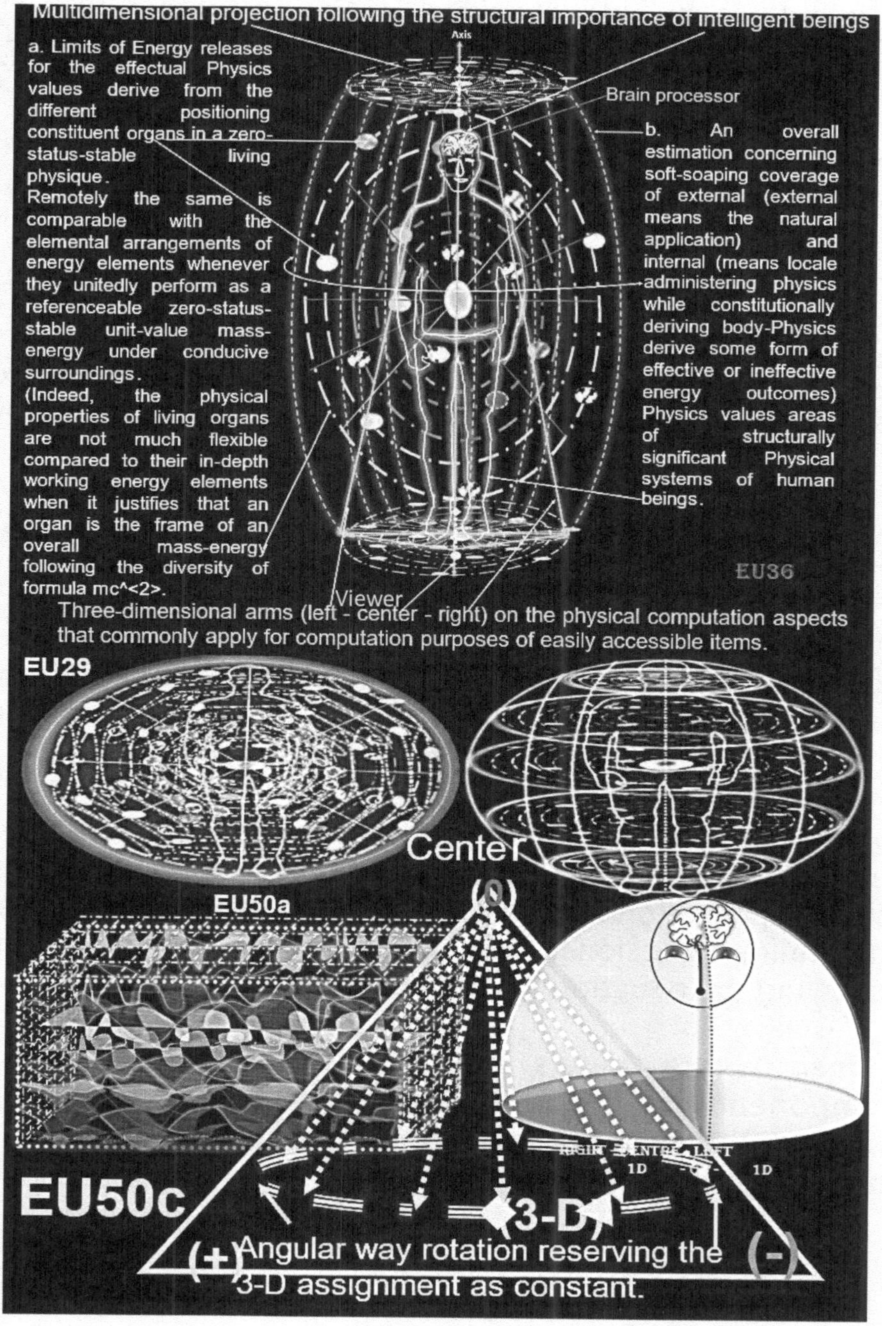
Multidimensional projection following the structural importance of intelligent beings
Axis
a. Limits of Energy releases for the effectual Physics values derive from the different positioning constituent organs in a zero-status-stable living physique.
Remotely the same is comparable with the elemental arrangements of energy elements whenever they unitedly perform as a referenceable zero-status-stable unit-value mass-energy under conducive surroundings.
(Indeed, the physical properties of living organs are not much flexible compared to their in-depth working energy elements when it justifies that an organ is the frame of an overall mass-energy following the diversity of formula mc^<2>.
Brain processor
b. An overall estimation concerning soft-soaping coverage of external (external means the natural application) and internal (means locale administering physics while constitutionally deriving body-Physics derive some form of effective or ineffective energy outcomes) Physics values areas of structurally significant Physical systems of human beings.
EU36
Viewer
Three-dimensional arms (left - center - right) on the physical computation aspects that commonly apply for computation purposes of easily accessible items.
EU29
Center
(0)
EU50a
RIGHT CENTRE LEFT
1D
1D
EU50c
(3-D)
(+)
Angular way rotation reserving the 3-D assignment as constant.
(-)

3. Multiple organs (*or at energy level may be assumed as operating energy-elements*) are positioned all around on different levels of orbital zones from close to the center to a certain distance due to standard gravitational attraction by the central soul (c) of the whole-body mass (m) and tend to work perfectly following the balancing application of internally deriving body physics, in coordination with surrounding atmospheric physics.

4. From the complex union of whole-body mass, this excerpt indicates an organ composed of multiple molecules, but the overall organ represents a mass-energy frame (mc^<2>).

In this sketch, as an organ, it (point#4) indicates that constitutionally settled molecules perform orbitally following centrally located some thinner amount of energy or the functional soul (m) for the organ (m) since molecules generally (or in healthy condition) systematically perform following internal and external physics values application in a conducive life-supporting inhabitable energy source.

→Drawing 50c (above) is a collection of a few earlier drawings to indicate the analogous position of 3D, from the simple application to a complex, more outstanding level of the multi-dimensional role of a complex living system following the equation E = mc^<2>.

Indeed, where there's mass (m) and motion-orienting component like (c) is involved, functionally deriving physics like Gravity and the roles of dimension is inevitable.

CHAPTER 16

From the standpoint of energy, some points concerning motion and motion-related items.

Initiation of the motion is due to the creation of Motional value (*that is, diversity deriving working essence*) from the motion element in the Supreme - Universe before forming the baby Universe and then the journey towards maturity to reach this modern stage. And the realms of multi-dimensions, which dilated their roles throughout the fathomless energy ocean of the 2nd-energy-belt, perfectly maintain inter-tangling relation for everything regardless of their status as mass-energy frames, objects directly or indirectly throughout the fathomless energy-Ocean in the 2nd-energy-belt.

[*Note: - The below points are relatively enriching clarifications from earlier books "Mythology, Philosophy, And Science. Book for the New Generations." Volume I & II]*

Herewith presenting some significant points and following the logic of intelligent beings' judgments, which rely on the restricted physical cognizance that delimits by some dimensional purview, it attempts to simplify the multi-directional roles of energies, which can implicate for understanding the energy levels work throughout the matured 2nd-energy-belt.

Indeed, Part 2 of this book (*Energy and the Complete Universe*) step-wise defined the sense of direction, dimension, and multi-dimensions, following the general conceptions of bio-intelligence beings. Yet, in short, I will requote some points from the earlier books for clarity on the subject and initiation at the introductory level, that is, the inception level of the baby Universe.

i. The 'Motion element' is one of the most fundamental energy elements at energy levels.

a. Everyone is familiar that the word-sense motion alerts a human physical system when a small or large structurally rational object moves or changes position following some distinguishable types like oscillation, rotation, linear, non-linear, etc., following measurable times.

So, for observable sense, words like motion, move, and movement indicate some forms of activities whatsoever virtually at rest, still, or at zero state.

Also, everyone will accept that, for some reason, many detectable objects or items turn to active with some form of motion upon internal or external enactment or similar in the same sense, or otherwise the items will remain physically at rest, zero-state, or with no activity at all.

b. But, from the viewpoint of energy, the motion-deriving motion element is the root cause of everything for any discernible and non-observable items like momentarily appearing unproductive mass energy *(here, 'unproductive mass energy' means so much lowest level of momentariness of appearance incapable of characterizing as a frame of 'mc' for the prospective quantum particle*), microparticles and massive object.

Since the motion element is only self-governing at any level of energy mass, it occupies eternal dignity as a soul wherever it participates as the observable and non-observable items.

ii. At the state-of-Absoluteness of great grandmother Supreme - Universe, heavenly silence efforts of the motion-element anticipated for mainspring from birth, childhood, and then gradually impel to reach towards maturity of the Universe, upon segregating into three sections as 1st-energy-belt, 2nd-energy-belt and finally the 3rd-energy-belt. (*The drawing EU51 is already defined in an earlier book, which indicates the Motional value's initiation at the most fundamental point with the anticipation for the mainspring of the baby Universe in the Supreme - Universe.*)

For the above clarification, I may utilize the same famous equation E = mc^2, and its augmented status as E = mc^<2>, which is described in a break-up manner in earlier chapters for various purposes when employed to convince the alternate positions of mass-energy, microparticles, and large objects.

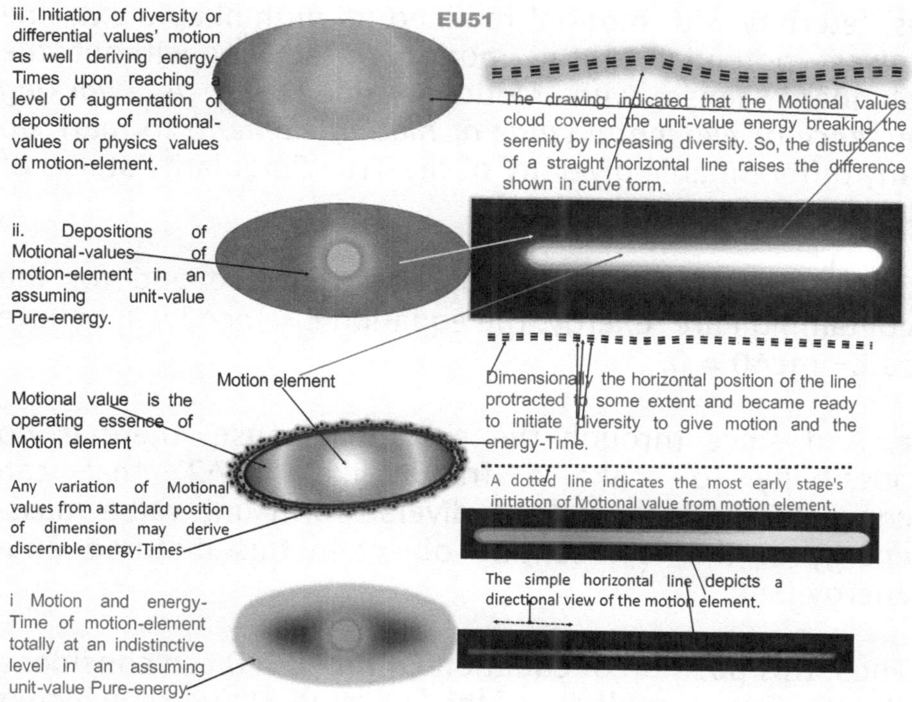

a. But for the Supreme - Universe and its State-of-Absoluteness or True singularity state of all realizable and unrealizable contained energies are at one and only one indifferentiable status that covers throughout the "no beginning and no end areas" since this sense mainly applies to objects and observable concept as a substitute or stopgap of physical nothingness by the intelligent human beings in this bio-Globe.

Otherwise, the existency of the great-grandmother Supreme - Universe is truly an eternal status, whatsoever the playful living minds search for the reality of nothingness in the fathomless diversifying energy -Ocean of the 2nd-energy-belt.

iii. Therefore, for the Supreme Universe, at the State-of-Absoluteness or True Singularity state, the motion element, which proprietorially occupied the heavenly blessing, that is, 'eternity with motion' retained its unthinkable extreme lowest potential to the motion or activity without the standard sense with Motional-value (*that is, working essence or elemental value of motion element*) to mark up any differentiable diversity of the True Singularity status of the Pure- energy.

So, at this stage of the Supreme Universe and for the containing Pure -energy, the equation E = mc^2 will behave as **E= mc^0 = 0**.

a. And since through the evidential sense, there is no possibility to make a frame like 'mc^2' that can conveniently apply for any diverse form unit-value mass-energy or the representing object in this matured 2nd-energy-belt.

Thus, this position of equation **E= mc^0 = 0** can consider a state of True singularity of the Supreme- Universe, which is applicable beyond the perceptible sense used by the intelligent beings in this bio-nature.

b. Based on that, even if someone intends to apply through the sense of microparticle levels, following the augmented formulaic status as E = mc^<2>, where 'E' acts as a state-of-stableness frame represented by the combining values as mc^<2>.

iv. Now, if the power of 'mc' means characterization by ^<2> reduces and replaces by the zero or 0, which mainly applies to observable sense for microparticles like (E0) or no-mass, rest-mass, etc^^{n}.

And for articles or things like **E = {(mc^<2>)0}** **peripheral,** which means an object internally active since there is the retainment of 'm' and 'c' with some power value that can undoubtedly raise diversity upon applying some stimulation or implication, but externally at rest or inactive, which convincingly denoted by the symbol 0.

Then notionally, the (E^{pure}) True state of Pure - energy also will not work for creating differentiation and for raising any Motional diversity that infectiously severed from Pure status and enhancing defilement since consciously the concept of 0 has taken status as nil, null, nothing, etc., that became significant for mathematical, physical, and many other ideological placements for diversity in the 2nd-energy-belt.

c. But in contrast, at the energy levels, everything is eternally active either to an acceptable lowest of potentiality (*this extreme lowest potentiality conveniently can be indicated by applying decimal values with positive and negative of span*) or highest to some status of kineticism in the 2nd-energy-belt.

Therefore, for the Pure-Energy of the Supreme - Universe, this 0 and one of its lowest decimal values, 0.0, is seemingly the best numeric option to define the formation of inconceivable the most-micro level Motional value or working essence for coverage of an extreme micro point (mc) or Pure state mass-energy unit.

v. Accordingly, the equation evaluates as (E^{Pure} = mc^<0.0>) for an innovative mass of Pure-energy unit in the great grandmother Supreme-Universe.

Considering the extreme lowest decimal value of the formula E^{Pure} = mc^<0.0>, it further may appraise as E^{Pure} = mc^< **(**-**)**0.0**(+)**> for application at the initiation of diversity before it infectiously widened incrementally to cover like the fetus areas for formation of the Universe in the mother Supreme-Universe (EU52).

[For a unit of Pure - energy[EU52] *of the Supreme - Universe, the fundamental formula "E = mc^2" applied as: -*

- *E^{Pure} = mc^< 0.0>*

And: -

- *E^{Pure} = mc^< (-)0.0(+)>*

An appraisal state for application at the initiation of diversity before it infectiously widened incrementally to cover, like fetus areas for the formation of the Universe in the mother Supreme - Universe .]

vi. Next, upon the formation of the Universe, it occupies all forms of diversity and uprising that modern intelligent species observe, analyze, and even imagine based on the limit of intellectual insights while residentially positioned in this beautiful biologically orienting life-supporting bio-Globe in this matured 2nd-energy-belt.

Gradually and finally, from the baby stage of the Universe, the Motional values of motion elements, along with the other assisting elemental values or physics values of energy elements, gradually organized the entire quantity of energies into three energy belts (EU53) that facilitated the Universe to get with enduring dignity.

A projection of the three distinctive energy belts of the Universe that finally evolved before progressively matured.

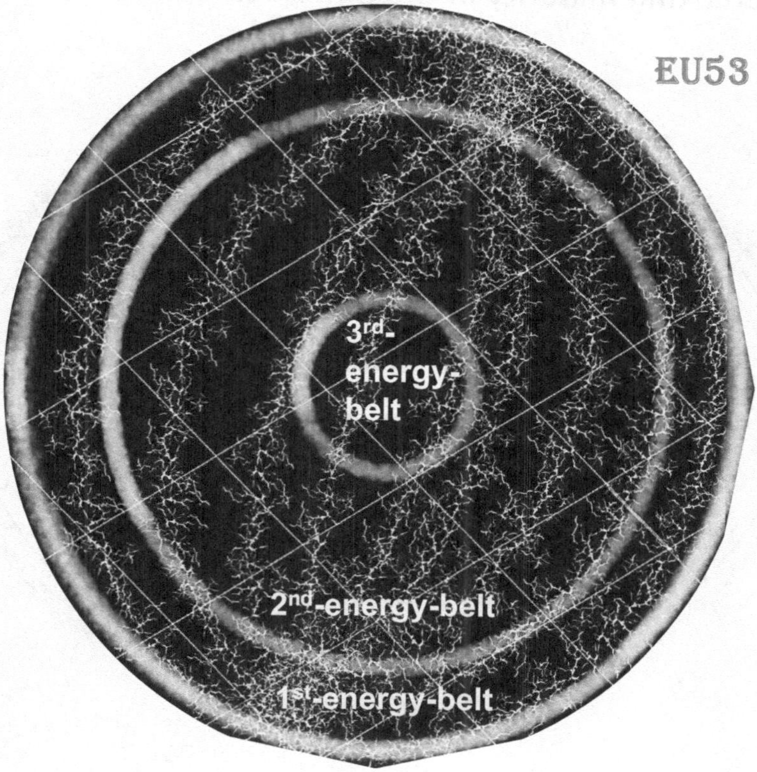

a. The 1st-energy-belt occupies many essential energy elements at their denser state and somewhat in an absolute ceasing state (*based on technicality, acceptable through the specific way designed perceptible sense*) before being locked within the boundary or belt that separates the 2nd-energy and the 3rd-energy-belt.

b. Gradually, the 2nd-energy-belt derived or filtered away the extreme finest form of energies or Purest- Energy in the 3rd-energy-belt before it locked and initiated to organize into different zones based on the diversified grades

(*Drawing EU54 is an assuming prospective configuration of some growing sections of energies in the developing 2nd-energy-belt indicated in EU53*) within the 1st-energy-belt and the 3rd-energy-belt, and progressively reached a state of everlasting maturity in the Universe.

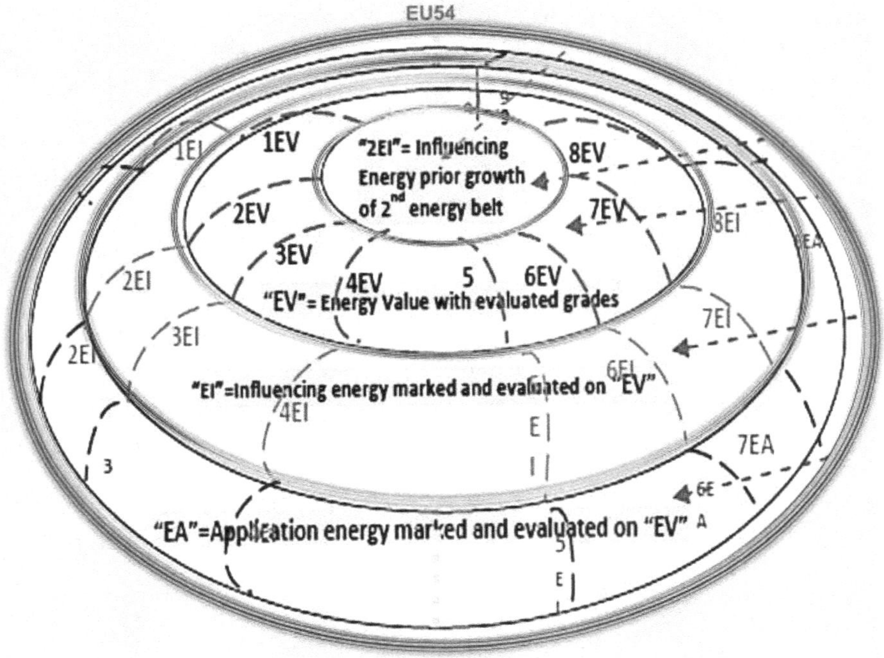

iv. Even though after the maturity of the Universe, the Motional values work of motion elements never ceased, its eternal activities are prominent in this 2nd-energy-belt that living beings consciously can understand every momentariness of life's movements, and even the non-living products have to follow the transformation rules since at energy levels, Motional values do not allow to stay at True permanency from an earlier status as an object or mass-energy frame. Indeed, in the perceptible sense, the word balancedness in place of permanency seems comfortable and appropriate for the observably considered long-lasting items.

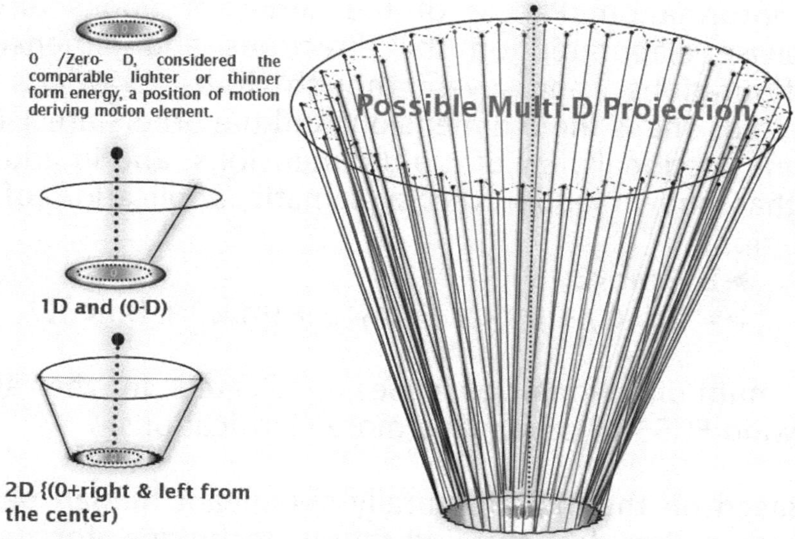

A sectional view of a multi-dimensionally performing matured unit-value mass-energy frame, ready to give a structurally definable object /matter /material upon application of internal and external physics, that is, Elemental-values of energy elements and externally supporting atmospheric physics (i.e., Twig + energy atmosphere).

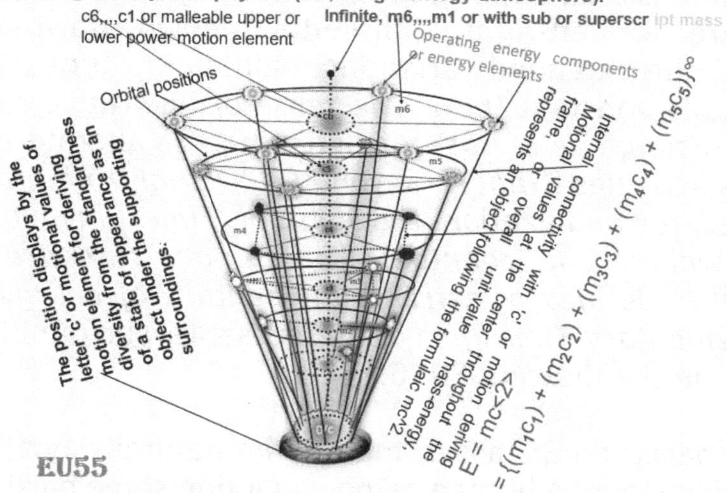

Orbits and elemental arrangements of energy elements interlinking with motion -element 'c'.

a. From the dimensional perspective, I think the symbol '0' is a perfect numeric value to define any dimensional role, including the energy level's multi-dimensional activities in the 2nd-energy-belt.

For information, Part 2 of the series of books already stepwise elaborated on the directions and dimensional configurations, and even though four- drawings are selected, one is the 0 dimension, and the others are 1-D, 2-D, one sectional view of multi-dimensions, and in addition to that shown about the systematic application of the equation: -

▶E = mc<2>

▶$= \{(m_1c_1) + (m_2c_2) + (m_3c_3) + (m_4c_4) + (m_5c_5)\}^{\infty}$,

for multi-directional purposes, indicated in the above drawing EU55 to ease many more clarifications.

v. Based on the 3D structurally significant human beings' activities, besides the judication technique for taking decisions of most direct and indirectly assessable items, which principally works as a self-bearing constant like 0 for two-dimensional limiting roles, following angular rotations that are like left-right, right-left, up-down, down-up and further the extended variations maintaining 0 or (c) as an invariable constant (*here, the sense of invariable constant is a presumptive status*) for detecting the altering physics values activities {(*that is, beyond 0+2D analyzing practices*) (*indeed, for analyzation, the sense of 'Time' play most vital role whether it countable based on standard time, unrealizable idea of extreme momentary of energy-Times or elementary-Times*)} in an assessable surrounding, as shown in the drawing EU56.

a. So, based on 3D assessments, the natural way affiliated body-physics of a human being can judge some partial roles through the delimiting sensing systems, whereas many inbuilt molecular functions and their constituent working essences or elemental activities of different energy elements mostly remain neglected while taking decisions by brain processor of intelligent beings.

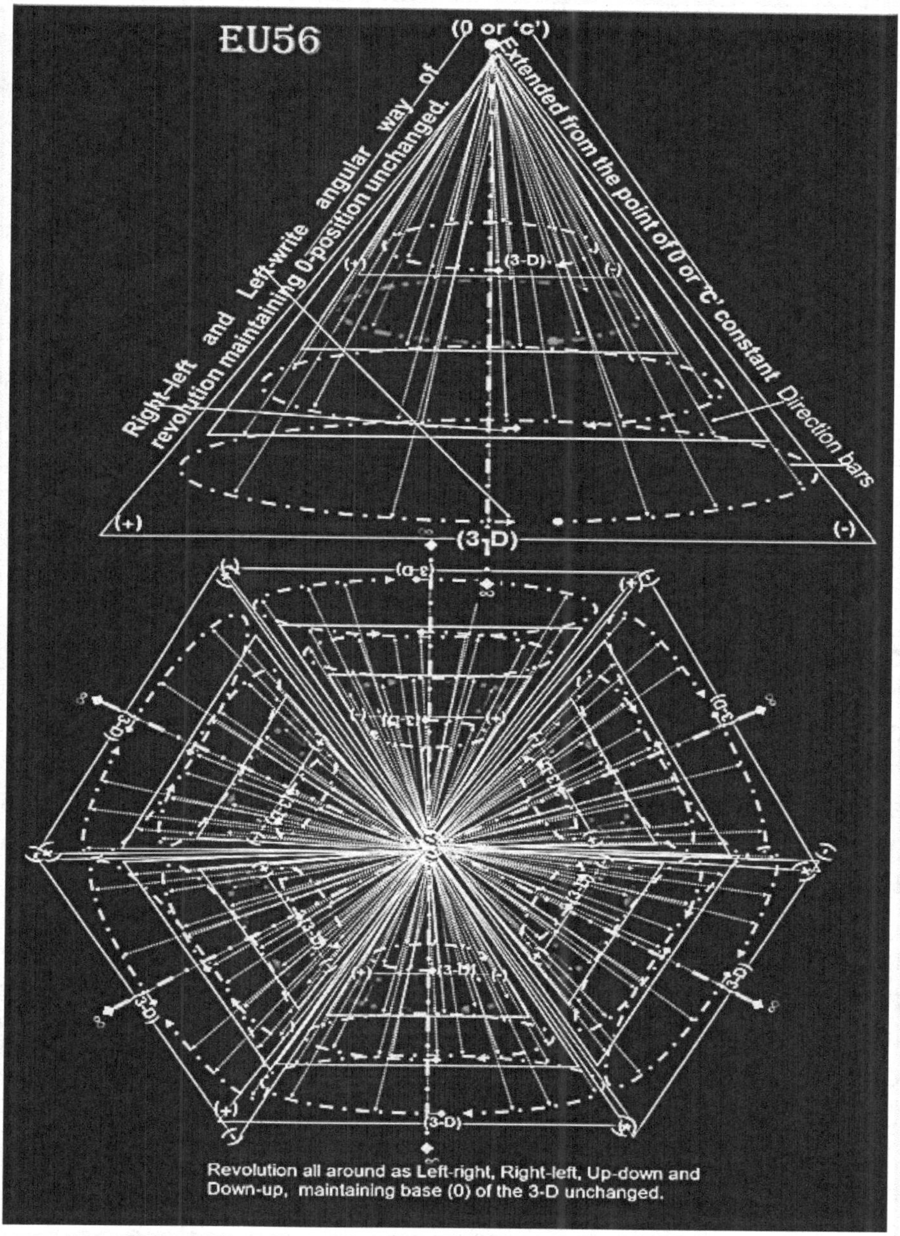

b. Logically, it may convey that deficiency is retained at the grassroots level of molecules when human beings judge /analyze something complying with some assessable items compared to the energy levels' multi-dimensional purview.

Therefore, from the above, it hints that on the aspects of universal multi-dimensional physics values roles, human beings have to hurdle for analysis and take confirmative decisions that fall beyond the limitations of the physical systems provided by the mother nature bio-Globe.

Accordingly, for computational purposes through the brain processor, intelligent biological beings found two very fundamental points that are 0 (*zero*) and ∞ (*infinite*) for an acting beginning and the endpoint when measurements perform following some near or distantly accessible coordinates for measurements, and anyhow if the resultant outcome become unsatisfactory to an individual or group strategy then decision fall to the perimeter of the symbol like infinite, dark zone or the senses applied by the word, unacceptable, unbelievable or unrealistic, etc.

c. Also, it says that intelligent beings, either of single or their united consensus, mostly decide on an average outcome for energy-level multi-dimensional activities beyond the limits of primarily the direct (so, the indirect access more than 3-D, always highlights as probable marks) analyzable physically appearing 3D constant, somewhat indicated in the drawing EU57.

Or, the same may clarify that any intellectual being, whether acting individually or through a collective consensus, often influences and determines outcomes related to multi-dimensional roles and energy activities that extend beyond the primary, directly assessable dimensions.

In this context, the physical 3D fact the living beings observe is merely the most immediate covering of a far more complex energy-level structure. The limits of what can be directly analyzed and generally curbed to 3D assessable space, which serves as the baseline.

However, energy activities and interactions occur in higher dimensions and remain as indirect accessibility. So, when the activities of higher dimensions are not fully perceptible by bio-intelligent beings, quit from traces, putting "probable marks," allowing advanced intelligence to infer and engage with them.

The drawing EU57 also subtly indicates that beyond the perceivable constant of 3D space, intelligent beings can tap into and influence multi-dimensional layers, shaping energy flows and outcomes in ways that transcend physical limitations.

It suggests a reality where consciousness, perception, and energy are not confined to traditional dimensions but are interconnected across a more flowing /runny, dynamic, and multi-dimensional framework.

In this way, intelligent beings act as bridges between dimensions, accessing a more comprehensive energy system that operates beyond the directly perceptible items and observable universes.

The collective actions of energies are not merely responses to the physical world but also shaping forces that influence the entire structure of the actual Universe across multiple dimensions.

So, it emphasizes the interplay between higher-dimensional energy systems and the limited role of intelligent species in some controlling outcomes across the limited dimensions. It also adds in-depth to indirect access and explores how intelligent species interact with the energies, which are beyond the observable limits but remain mysteries since it is methodically unanalyzable.

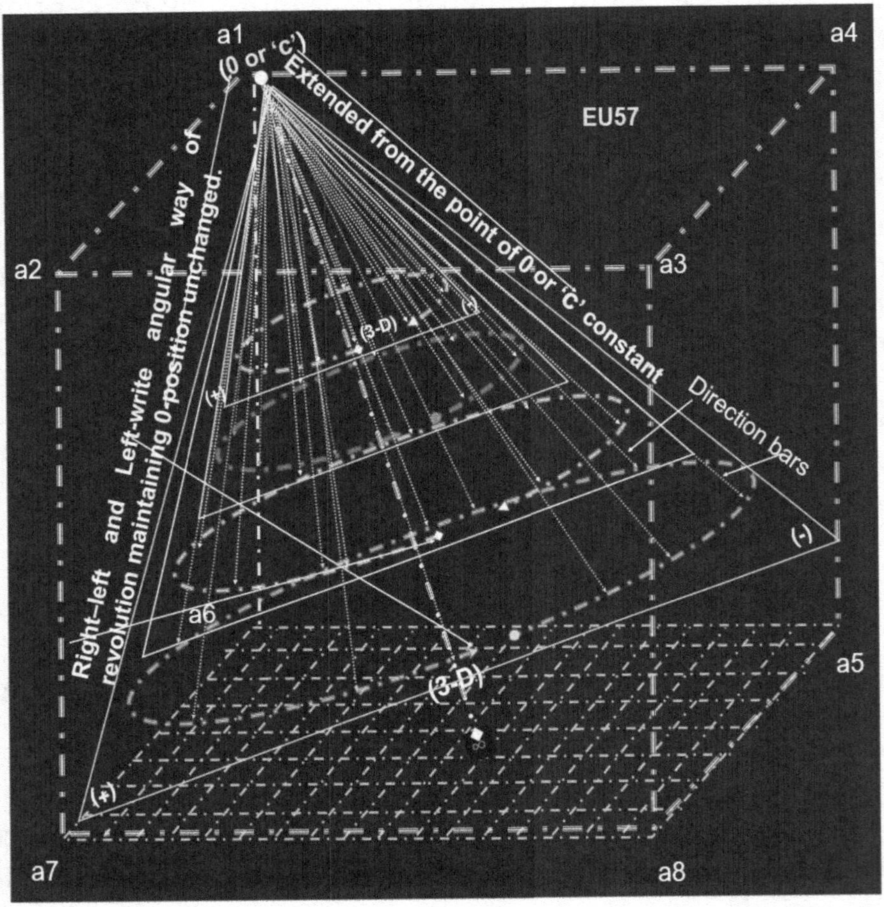

d. Drawing EU58 presented with a rectangular shape occupying some quantity mass form energy, assuming coverage of eight (*a1 to a8*) accessible points for analyzing the all containing regardless of most microparticles, large objects, and others. So, from the viewpoint of energy, the outcome of the eight-point analysis will be the energy level's overall result.

(In practice, the individualism of each human being performs somewhat like the natural way developed 3-D measuring mechanism. So, it acts as a mathematical constant or 'c' for all assessable differential analysis.

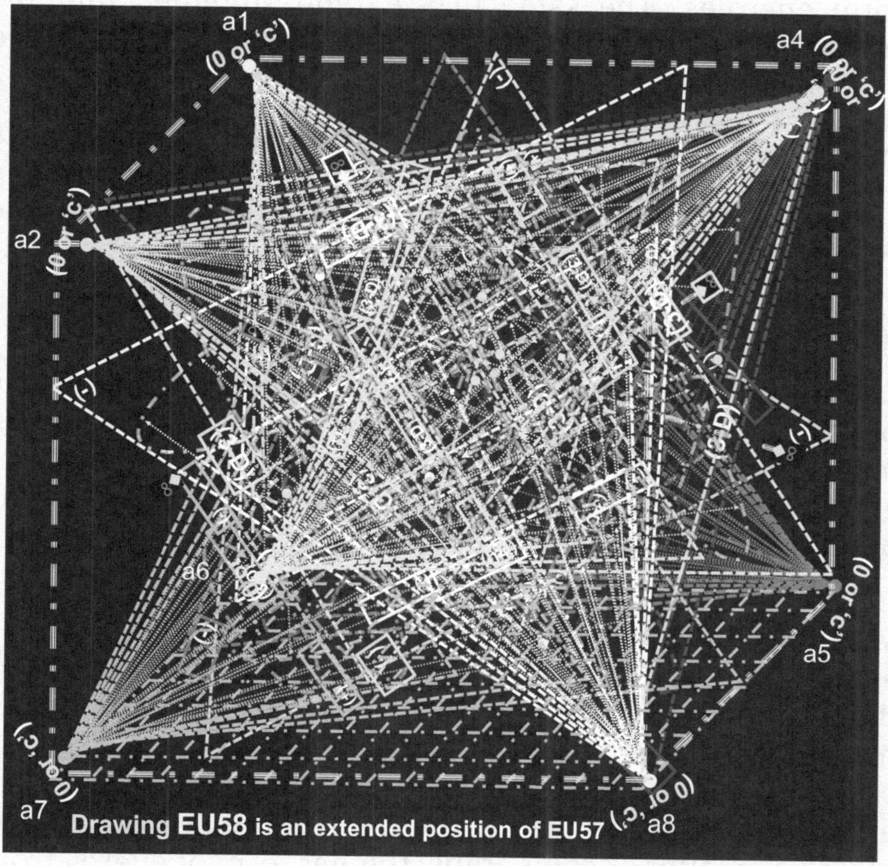

Drawing **EU58** is an extended position of EU57

Thereon, for extra-dimensional purposes, it needs to adjoin indirectly over and above the natural or direct gift that is 3D, then to make 4-D, 5-D, and so on up to the multi-D range that will cover all dimensional diversity for the entire fathomless energy-Ocean of the 2nd-energy-belt, whether assessable or realizable directly through the self-bearing brain processor of intelligent beings or indirectly by any probable, extra means.)

e. Ideologically, the coverage of a rectangle {(drawing EU58) (*basically, a rectangle justified as easily accessible 2-D, like length and width, following the point (0) function as a base for analysis*)} is small compared to the vast areas of Energy-Ocean and microparticles, massive objects, and other mass form energies. The same may extend adjoining multiple rectangles for covering large areas of energy Ocean in the 2nd-energy-belt.

And so, at this moment, multiple unified rectangles are conveniently indicated by using a cylindrical shape encapsulated with some mass of energy since the cylindrical (*drawing EU59*) shape performs comfortably assessable 3-D, like length, width, and height from a point (*0 /base*) of analysis

f. Since the measurement practices mainly through the physical 3-D analysis, it stands as the natural reality to many intelligent beings, and the question of acceptance about energy levels' multi-D (*multi-dimension*) role remains paradoxical, unbelievable, probable, fictitious, and likewise status.

vi. Following the constant 3-D physical reality, now a simple and minimum convincible way tries to clarify the multi-dimensional systematic roles of energies throughout the matured 2nd-energy-belt. (*Drawing EU59*)

a. In the matured Universe, the 2nd-energy-belt occupies a massive amount of comparably thinner form of energy from the rest of other diverse grades, parametrically considerably lower and higher degrees of denser forms of energy besides multiple types of objects like matter, material, materialistic items of their most micro to large states and their progenetic energy products.

This thinnest form of energy in this matured 2nd-energy-belt can be dignified as ostensive Purest-Energy, proprietorially unique since it occupies extreme speed and neutrality in all aspects covered throughout the 2nd-energy-belt.

(*Concerning the ostensive Purest-energy and its importance, clarified in many places in all books.*)

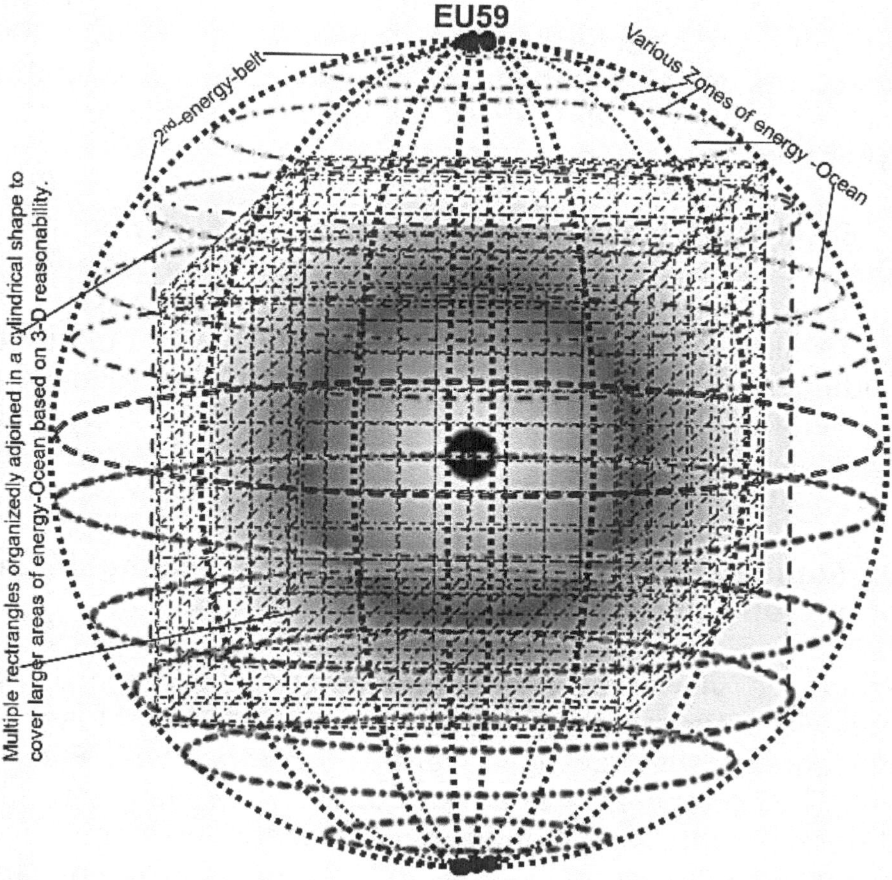

Due to uniqueness in aspects of the extremity of speed, neutrality, and the volumetric concept, the ostensive Purest-energy unquestionably became the unalterable soul for the 2nd-energy-belt, so it functions like a mathematical constant, and as well as the base point or '0, zero' in all aspects of physical reality that follow mainly the 3D computation and analysis.

b. Also, for biological beings and many other alien species, it (*ostensive Purest- energy*) acts as an end-point like an infinite for covering the multi-D journey following the limits of the 3-D principality, maintaining self-bearing constant like the 0 and altering two-dimensional limiting roles following angular and cyclic maneuvering, keeping a fixed point for revolutions like the left-right, right-left, up-down, down-up, and even further the extended variations, whether it accepts as reasonability of countable or uncountable moment through the physics values realizable event/s (*shown in the earlier drawing EU56*).

c. Besides, it may requote that 'from the viewpoint of energy' and following the Universal law of energy transformation, there is no possibility of increasing or decreasing the quantity of energy occupied by the matured 2nd-energy-belt. Thus, all energy products must follow the conversion process, whether perceptibly appearing cyclically, recycling, or entirely vanishing from the discernible limits.

vii. Some evaluative explanation about the entanglement of microparticles.

From the viewpoint of entanglement of microparticles, herewith an attempt to merge the ideas and explanations presented in the drawing EU4 and EU30 of the book, 'Energy and The Complete Universe' Part 3, explained linguistically different fontal exposition of alphabetic elements and proposed enhancing its links to unimaginable far distant quantum particles by employing some of multiple linguistic alphabets /elements and their different typeface, those accepted for literature purpose by the human intelligent in this bio-Globe of the 2nd-energy-belt with the Drawing EU60.

a. The Predictable reason about the commencement of language!

Since no one knows the actual beginning or appearance of the human species (*obviously, everything appearance and nonappearance of as living physical status, non-physical, non-existent or fantasy items to human species are the derivatives due to diversifying efforts of energy-God*), it may be reasonable to predict that the true origin of language and improved way of expression through the script or written system with visible marks initiated gradually towards the evolution of living like the primordial human species.

→Those (*ancient human species*) who learned natural ways following the uniqueness of structural or constructional abilities and ensuing the roles of some significant internal and externally particularized organs like hands, legs, visualizing eyes, hearing ears, arrangement of vocal cords, brain processor, and some other factors that cooperatively work out to find better techniques of expression and communication in a place of survival or conducive atmosphere, which subsequently allowed to initiate enhancement and reach this modern physical way of expression by using alphabetic elements and through machine use languages or technically deriving coded forms communicating mechanisms.

→ *{(Here, from the perspective of the sense of entanglement, living intelligent species of an inhabitable nature become closely entangled with the items to their physical proximity due to the regular interaction for survival necessities. But, when some predictable things are explored through the brain processor and transformed into notional reality, the distant entities turn to close entanglement for acceptance.)*

(Further, from the viewpoint of energy, modern remote transmission techniques of information became closely entangled for physical realization. So, the methods for communication during the primordial eras' human species became retarded or distantly related since mindset or mental dispositional attributions tend to erode following insignificance of utilization while existing in the contemporary technological culture.)}

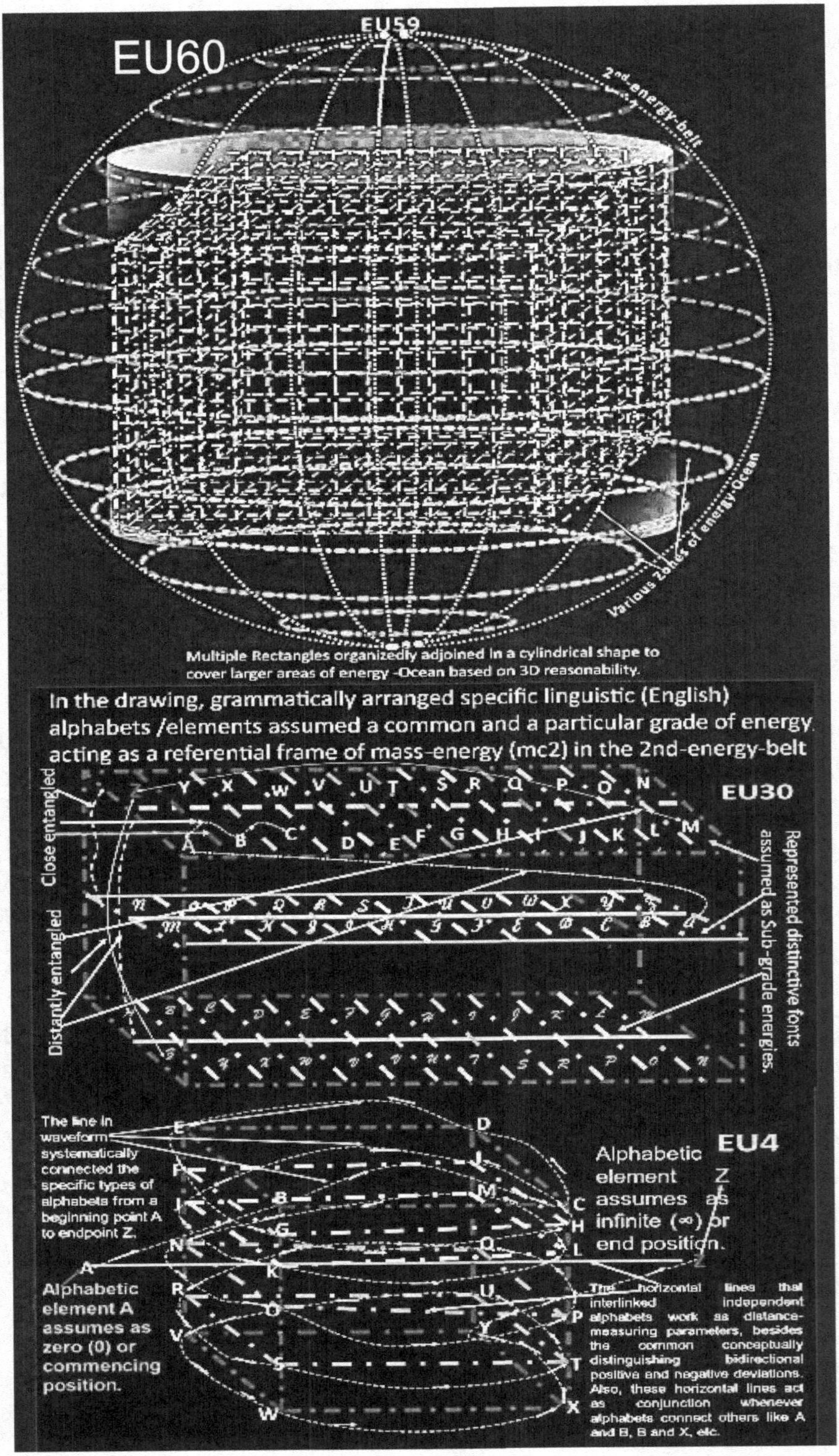
EU59
EU60
2nd-energy-belt
Various Zones of energy-Ocean
Multiple Rectangles organizedly adjoined in a cylindrical shape to cover larger areas of energy -Ocean based on 3D reasonability.
In the drawing, grammatically arranged specific linguistic (English) alphabets /elements assumed a common and a particular grade of energy acting as a referential frame of mass-energy (mc2) in the 2nd-energy-belt
EU30
Close entangled
Distantly entangled
Represented distinctive fonts assumed as Sub-grade energies.
EU4
The line in waveform systematically connected the specific types of alphabets from a beginning point A to endpoint Z.
Alphabetic element Z assumes as infinite (∞) or end position.
Alphabetic element A assumes as zero (0) or commencing position.
The horizontal lines that interlinked independent alphabets work as distance-measuring parameters, besides the common conceptually distinguishing bidirectional positive and negative deviations. Also, these horizontal lines act as conjunction whenever alphabets connect others like A and B, B and X, etc.

→Indeed, for all those communication techniques, indubitably involvements of energies and the different ways of atmosphere and atmospheric physics that momentarily administered to reach this modern improved application of linguistic signals that can deliver nearer and even far-far distant in the form of audio, visual, combining of audiovisual and more advance, directly to inbuilt setup sensing organs related communication expressions through natural as well mechanical means.

→ {(*Here may quote, combining efforts of constitutionally sets sensing organs organize and ensure the possible security of an entire living physical system while evolving in a conducive atmosphere.*

Accordingly, some sensing organs developed on a priority basis from others in the bodily systems of living species. So, the human species became one of the best creations of Mother Nature's bio-Globe and got dominant features compared to many other creations in the aspects of specialized sensing organs and physical constitution in this bio-Globe.)

(*Here is with some quantum fantasy assuming that the doctor declared a cat dead for some reason following the physical inspection.*

In a summarizing manner, it conveys that whatsoever the biologically stated reason (that is, disease, infection, circumstantial incident, accident, etc.) *for the death of the cat, at the energy level's technical extract is that the body physics of the cat became incapable of checking with a limited capacity of inbuilt organs or union of molecules and manage to maintain balance upon application of external energy physics.*

But, on the aspects of affluent life-orienting energy in a place of dwelling, which regulates the life of the living cat, death should not happen if surrounding atmospheric physics and constitutionally deriving body physics of the cat could manage healthy relations.

Here, applying the word sense diplomacy on the aspects of creations and destruction, mother nature bio-Globe played a vital role since she has to abide by the universal laws of diversity in the 2nd-energy-belt, whatsoever she could earn a specific kind of life-orienting energy, that is, bio-grade energy specialized for biological livings.)}

b. Since energies are the whole and sole of everything, languages, linguistic scripts, or alphabetic elements with different typefaces are the parts of energy signals conveyed through the inbuilt organs specialized for expression.

→Indeed, human beings act as natural performers for constructing different ways of expression methods like gesturing, audio, visual, script, and coded, machine-generated signals for improved communications.

→Therefore, some languages with different typefaces and linguistic alphabets or elements are suitably employed to illustrate the entanglements of microparticles formed in various rectangle shapes containing more or less (but not equal) from particular classes of energy to near and far distant microparticles *(indicated in EU60)*.

c. For examples of spooky connections between near and distant mass-energy frames or microparticles, conveniently explain by means of various alphabetic elements.

Technically, specific (*means proprietorially acceptable as one of its kind*) microparticles can maintain perfect communication relations provided proprietorial typefaces, like linguistic alphabets, and distinct fonts used for different placement and expression in proper word forms to retain firm and meaningful status for their objective.

{(Here, linguistic elements presumed similar to the microparticles, and the sense 'typefaces' are mainly used to indicate the distinctiveness of individuality of microparticles on the aspects of recognitions based on the parcel they occupy as energy form mc^<2> unit.)

(Further, to make an English meaningful word and sentence, placement with capital and small letters must abide by the grammatically provided standard rules that make it easy for common acceptance.

However, the sudden inclusion of any unfamiliar or ancient script to make an English meaningful word and sentence. Then the meaning of words and sentences will be unrealizable and unacceptable to most users, even though all those are in weird relation to the aspects and objectives of linguistic expressions.)

The drawing EU61 attempted to indicate how orderly entangled one type of linguistic elements pick up from their close to distance for making meaningful words, and then following the grammatical rules, words are places to make a sentence for sensible expression.

Besides, some different scripts like, 'Egyptian hieroglyphs, 'Edwardian Script ITC, 'SoupLeaf' "KG Flavor And Frames Seven, along with modern English letters, are used to indicate difficulties in understanding the meaning of words and sentences if unknown, even though they have a common ***purpose*** *and application, that is, vocal and visual ways of expression.*

*It (*that is, for literature purposes*) is somewhat similar and reasonable to say that proprietorially, one type of microparticle constitutionally sets more or less of the same class of energy to perform some delimiting degree of performance for a specific* ***purpose****.)}*

d. From the viewpoint of energy, it may convey that alteration may happen to any micro-particles with the momentariness of energy physics following the conversion laws of the energy.

→So, the alphabetic fonts also innovatively improved from the prehistoric proto-writing systems following the diversifying roles in the 2nd-energy-belt, and since energies never retained a standstill in any place of locale.

[Following points marked in the drawing EU61: -

→i. Besides, some different scripts like, 'Egyptian hieroglyphs, 'Edwardian Script ITC, 'SoupLeaf' "KG Flavor and Frames Seven, along with commonly use English letters, are used to indicate difficulties in understanding the meaning of words and sentences if unknown, even though they have a common purpose and application, that is, vocal and visual ways of expression.

→ii. Letters picked up from different placements following the standard grammatical rules for making meaningful words and a sentence,

→iii. Linguistic elements in the capital and small form are orderly entangled from close to distant ones from A to Z and 'a -to- z.

→iv. This drawing indicates the possibilities of microparticles' entanglement with the near and distant ones for a particular purpose. Like the independent capital and small linguistic elements randomly pick up from orderly arrangement for uses to express words and sentences.]

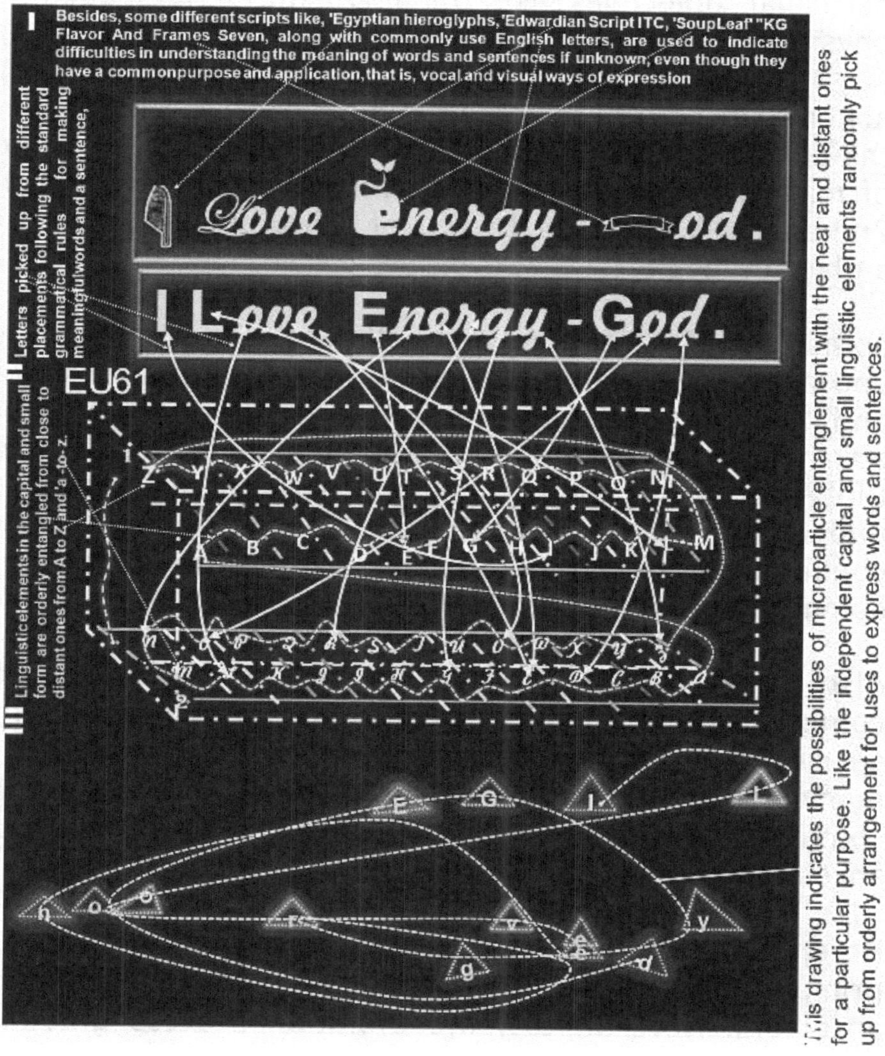

viii. In the drawing EU62, presume that various linguistic scripts and their specific types are independently blocked in near and distant places of rectangles to indicate more considerable coverage of 'energy-Ocean' in the 2nd-energy-belt.

Or, in place of different groups of linguistic elements, it may consider that proprietorially specific groups of microparticles settled in different rectangles that cover some mass-energy areas in the 'energy -Ocean' and perform particular objectives like the linguistic scripts used for physically producing communication, expressions, etc.

a. Remember, from the standpoint of energy and following the constructional constancy of intelligent beings, groups of microparticles cannot be perfectly particularized in the 2nd-energy-belt on the aspects of behavioral recognition in a place of locale.

Something like assuming an intelligent alien species grown and adapted in a different class of biological atmosphere and capable of communicating and expressing more advanced than the contemporary bio-beings in this bio-Globe, obviously both intelligent species will hurdle to understand their language, even though both communication techniques and expression uses for a common purpose.

Also, for a close example, look at the expression and communication techniques of other primate biological species in this bio-Globe, which is difficult to decipher to many intelligent beings even though those primates are the common cousins and have evolved from a common ancestor.

Here, it is clear and reasonable to say that all the modern primates once were closely entangled with their common ancestors. Nevertheless, the evolution process gradually improved the human species, deriving away the closeness from other primates.

b. So, based on the 3-D judication of intelligent beings, the microparticles that emerge in this natural atmosphere may not perfectly match a distant locale. However, considering the point of objectives, they are one, and somewhat like various linguistic elements, scripts, pronunciation, etc., are physically unlike. But all have a common intention *(here, computationally, the word sense 'common intention' gives an average outcome that generally covers many known and* *unknown factors*) to solve better ways of physical expression.

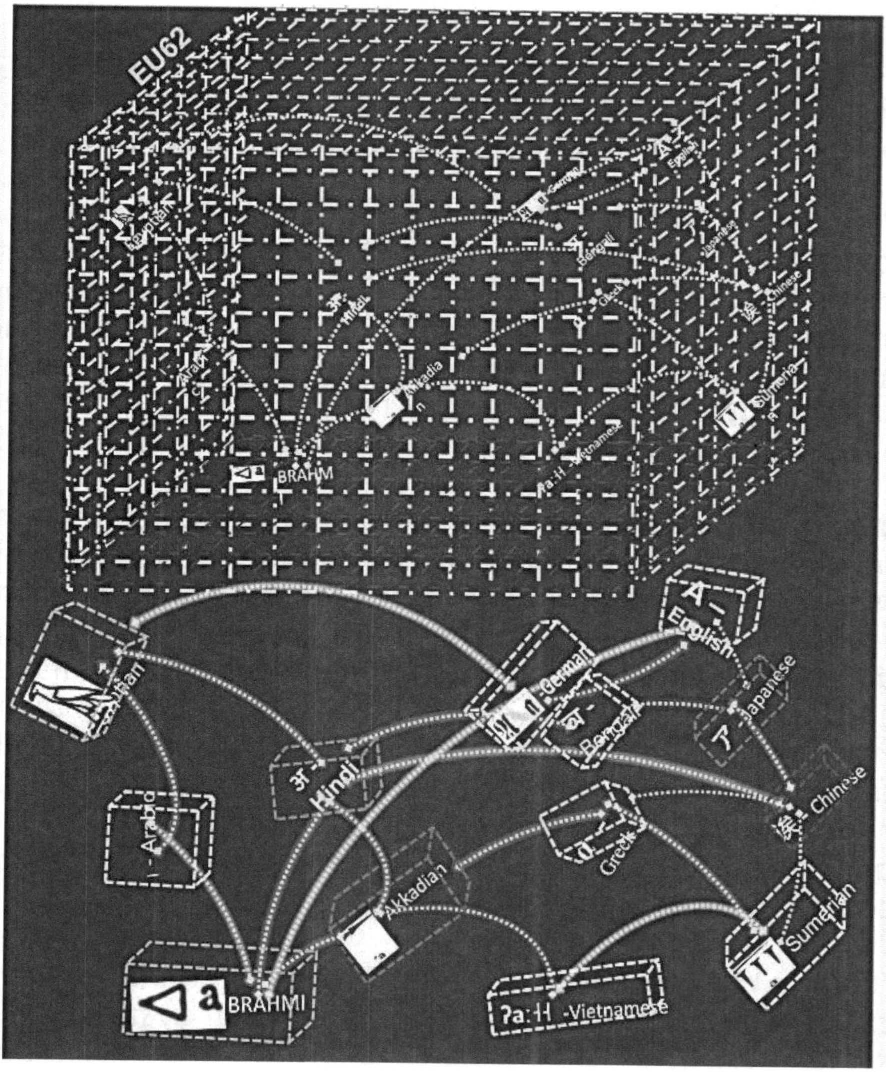

c. One thing to remember from earlier quotes is that the ostensive Purest- energy is the True Soul and proprietorially has unique characteristics on the point of speediness and neutralism for all communication purposes generated by the energy products, regardless of their individualisms *(here individualism means 'self-bearing constancy' following the quality and quantity of mass-energy (mc^<2>)*

representing as a unit-value frame of something), positioned nearer and distant from each other in the 2nd-energy-belt.

→Besides the eternity of effective constancy for everything, the 'ostensive Purest-energy' also directly or indirectly performs as the fastest universal carrier.

Principally can justify following the stance of energy, the 'ostensive Purest-energy' also performs as the fastest universal carrier or transporter for everything, including information signals for the most micro-level particles and micro-state mass-energy frame/s, provided the information remains unaffected due to the reasoning of internal and external physics values (*energy-Physics*) activities.

d. Indeed, for mechanical means or non-natural ways of transporting information using communication signals (*mainly analog and digital*), which are mostly invented by modern bio-intelligent beings, manipulating many differential constancy of proprietorially appropriate matter and material products (*material products are for making mechanical components*) following their total energy levels trait (*that is, qualitatively representing frame as a mass-energy unit/s*) that acts with lots of impediments, so the question of covering ranges of best quality of carrier for communicating signals beyond some assessable limits.

(Comparing the energy level's transportation, modern technologically developed communication systems give some desirable satisfaction following the delimiting requirements of intelligent human species.)

d. It is reasonable to say that transporting the information generated by the integrant physical systems of living and non-living in different locales through some available carriers. But at energy levels, those practices for the living system are somewhat identical to the energy levels' thinner class of energy, which follow some specific paths, networking channels or carriers represented as a small or massive cloud of energy mass, act somewhat faster compared to many energy- product's speedy features.

{(Remember, for all the cases, the ostensive Purest- energy acts as a True constant regardless of any mass-energy or energy product incessantly continuing, interrupting, or stopping a considerable journey.

Also, one technical point to quote is that whenever some quantity of energy is positioned as zero-status-stableness regardless of the primary stage's mass-energy frames or advanced mass-energy frames characterized as objects, their internal physics are very much balanced in coordination with the external atmospheric physics wherever they reside or travel.)

(Besides, one thing may recite that the ostensive Purest-energy inherently occupied its isolated fastest property that neutrally performs as eternal constancy for everything in this 2nd-energy-belt, whatsoever the intelligent species tend to analyze the perceivable diversity following the delimiting dimensional mastery and revolve this way or that way within the computational barriers like 0 and infinite.)}

e. By the by, In the present classical eras, the above seems an unbelievable or supernatural explanation to many intelligent beings.

In this regard, for clarification, I may emphasize with some examples about the development of communication techniques from the prehistoric physical appearance communication to modern eras paraphernalia transmission techniques that relied on some natural, non-natural, and practical way of describing means, which is far-far distant from energy levels' communication that realized and divinely applied by many physically appeared as God and Goddess or True-holy preachers for sacred purposes for biologically positioned intelligent species in this bio-Globe.

(Preach like 'God is one and only one' for all his creations. The soul has an eternal identity, whether someone represents a biological life in this bio-Globe or a distant inhabitable planet set to enjoy non-biological life.

So, following the energy level's narration, I may quote, 'Energy-God' is only the True position for all energy products, regardless of the partial way of presentation diversifying disposition by individual living beings or their group forms represented as socio-religious Gods, influenced and exaggerated due to materialistic acquisitiveness giving too much importance of presentness as appearance. Which at some point led to a call for misery to mother nature's exquisite creations and even may be directed towards ultimate destruction of the habitable property of this bio-Globe.)

f. ►Example1 with a drawing EU63: -

Something like comparable, countable many walking steps or assumed speed on the Times of a small ant with six legs and the instantaneous one long footstep of a human being. Assuming that a weary ant (A1) has to pass quickly, its bodily-carrying message or information to his mate (A2), who is eagerly waiting at one unit footstep's distance (*that is, some discernible measurement coverage*) from the human being.

Based on the general observable perspective, I think everyone will accept comparably many measuring steps or networking units for communication by an ant, which will equal the one physical unit of stepping Time or assume networking unit for a human being.

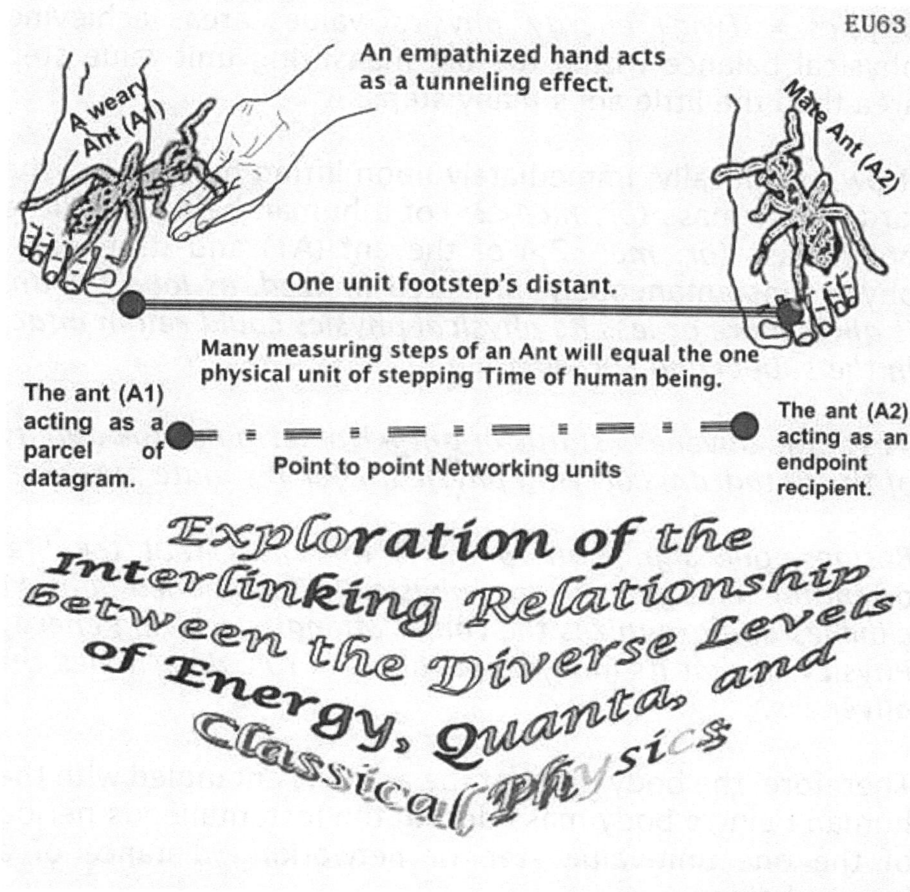

Concerning the technicality of open passage transporting media for the small body-mass (mc^<2>) of the unit-value ant and its inbuilt energy-physics, it has to suffer a lot due to atmospheric physics while covering the one physical unit of the step of a human beings' large mass (mc^<2>).

f1. Clarification: -

Supposedly, an empathized man realized the stepping struggles of the hurried ant, so he lifted it and instantaneously transported the ant (A1) to the target destination or meeting point of the mate-Ant (A2).

From an energetic perspective, that is, occupancy of the mass-energy frame, analogously to the ant (A1), the human being covers large atmospheric physics (*i.e., atmospheric physics = 'Twig+ energy' physics*) values areas achieving physical balance within the big measuring unit-value step area than the little ant's many steps.

{Now, technically, immediately upon lifting by applying the large body mass (*or, mc^<2>)* of a human being, the little body mass (*or, mc^<2>*) of the ant (A1) and its energy-physics instantaneously mitigated (*indeed, as long the ant is alive, more or less its physical physics could retain intact in the supporting surrounding.*)

(Also, *the aliveness status of ant-A1 determines the quality of the datagram carrying physically for his mate ant-A2.*

Besides, one significant point is to reflect that the life-orienting energy of the physically positioned ant-A1 clinches and organizes the constitutionally deriving energy Physics so that its (ant-A1) parametric role determines the aliveness.)}

Therefore, the body mass of the ant (A1) entangled with the human being's body mass during the instantaneous period of the one unit-value step or networking distance of a human being.

Here, from the quantum perspective of energy, the body mass of the human being instantly serves as a **tunneling deed** for the many steps of networking coverage of the wearied ant (A1).

g. Moreover, drawing 'EU63a' clarifies the interlinking relationship and performance of energy, quanta, and classical physics from their respective positions and roles.

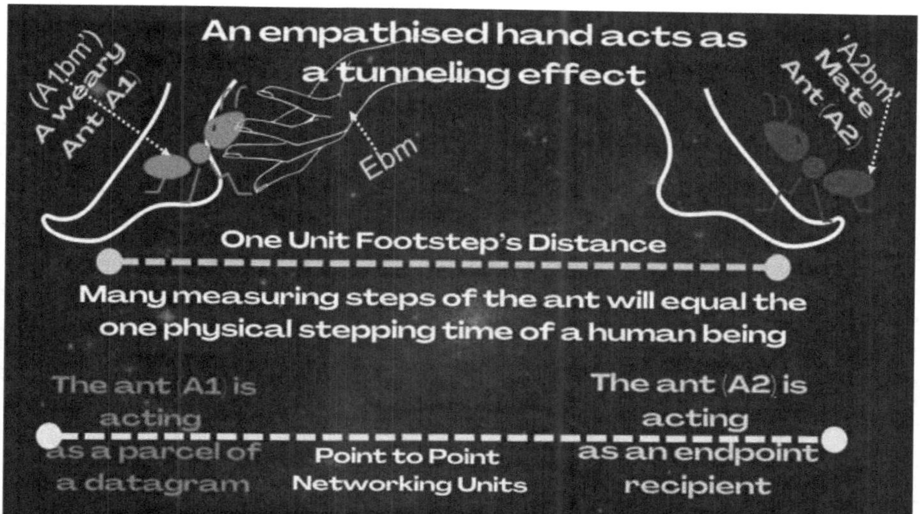

'EU63a' clarifies the interlinking relationship and performance of energy, quanta, and classical physics from their respective positions and roles.

Here, assuming as: -

The characteristically represented body mass of empathized man = Ebm

And the body mass of the Ant 'A1' and 'A2' = Respectively as 'A1bm' and 'A2bm'

> **i.** Following classical physics, I think it is reasonable to say that all the body mass as 'Ebm, 'A1bm' and 'A2bm are standard in a parametrically acceptable natural atmosphere since there are no unusual external physics (*classical*) values implications like temperature, pressure, resistance, etc., on their respective status.
>
> **ii.** But, from the standpoint of energy, energy Physics is consistently active with some form of loss and gain to maintain the parametric balance of the respective body mass as a state of their appearance as 'Ebm, 'A1bm, and 'A2bm.

iii. From the quantum viewpoint, if all three positions (*Ebm, A1bm, and A2bm*) of the respective body mass represent comparable positions of microparticles as mass-energy frames, then energy occupied by the Ebm will be much higher than the 'A1bm' and 'A2bm.

►Accordingly, the weary Ant or the representing parcel as a datagram with body mass 'A1bm' will indicate unbearable energy loss from the standard potential capability for some reason or due to excessive application of classical physics like load, kineticism, etc., and other related consequences.

Therefore, the excessive application of classical physics led to the utilization of unendurable energy physics from the body mass of 'A1bm' and eventually reached an unbalancing status on the point of reasonable recovery on 'energy-Times' of the 'A1bm' (*here energy-Times means inbuilt biological Times of the body mass A1bm*) in the surrounding atmospheric Physics.

►Indeed, considering the aspects of energy physics, Ebm acts as a tunneling deed while lifting and transporting the microparticle (body mass) A1bm, which is the most-micro amount loss and negligible since it is fast recoverable to 'Ebm.

{(Further, based on the prevailing classical concept concerning the loss and gain ratio, analogous to energy levels' functionalities of the empathizer man (Ebm) who performed as a logistic for the weary Ant (A1bm), is balanced.

Since, for those micro levels' activities, conceptionally seems one-way losses occurred due to the logistical performance by the 'Ebm' yet, his compassionate sympathetic role for the weary Ant, on the other hand, ***celestially energizes*** *his (Ebm) inner system for that moment of helping works.*

(Here, the word -sense ***'celestially energizes'*** *represents a socio-culturally emanating heavenly feeling influenced favorably by some imperceptible gain factors.)}*

g1. Now analogically, concerning the physical aspects of 'Ebm, A1bm, and A2bm, all three positions are independent in the point of energy, which occupied as mass form energy as well as the perspective of energy levels some dimensional coverage.

(*From the standpoint of energy, one thing is to quote in short that the enclosed amount of energy in a mass-energy frame determines the performable dimensional roles.*)

Since all of the body masses as Ebm, A1bm, and A2bm' follow the standardness of E=mc^<2>, they must able to generate some levels of Gravity because the body mass encompassed with mass (m), and motion deriving element (c) that have the potential to derive all internal energy physics including the Gravity.

But, on physical aspects of biological living, classically defined Gravity may not be acceptable to many intelligent beings and may choose to argue and seek alternative explanations.

{(*Since the* functionality as living *and orientation of life for the sympathizer being and the weary ant-A1 are from a typical grade of energy, that is, bio-energy. So, the conditionally deriving bio-sociological emotional roles* (something like, 'we are one' feeling*) from empathized man acts as kindness to weary Ant.*)}

→Thus, I find the word sense '**emotion**' (***emotion*** *means earthly /temporally /bodily deriving motion*) is one of the traditionally used spiritual words, suitable to imply physically generating Gravity or attraction-like roles (*that gives internal feelings like love, liking, alluring, desire, etc.,* *socioculturally applying nicety sense*) from a living body mass.

→Anti-gravitate effects

But noticeably, following the built-in 3D formal judgment procedures for living beings and many other alive physiques (*body mass*), the innerly deriving actual Gravity or attraction feelings work in contradicting manners, which may present by using the sense of anti-gravitate effects and may be meant by words and approaches with actions like repulsive, dislike, hatred, and so on, socioculturally recognizing oppositive of gravity-related affect.

This contradictory effect of gravitating or attraction feeling following Gravitational roles is due to the involvements of many realizable and inaccessible criteria, causal factors play inside one's body system, besides many other living body masses, matter, materialistically orienting socio-cultural developments, etc., and possibly remote or alien influencing signals that divert many original energies level standard works delivering with anti-gravitate countenances.

h. Example2: -

During the prehistoric eras, human beings' physical appearance was essential for reliable transmission or communication.

Whereas, for the primordial human outlook, techniques of talking or communicating by emitting sounds of some frequency levels by the plant world were contrasting in acknowledging. So, during the olden days, it justified that the natural methods of communication by the plant world were somewhat unrealistic and unbelievable.

In fact, from the viewpoint of diverse grades (*assuming parametric low and high degrees with differentiable quality*) of energies in the 2nd-energy-belt, consistent energy levels' changes invite slow and gradual evolutions in all structural and proprietorial values that appear to human observable senses while time-frame appearance (*that is the period of begin as living life to the cessation of adhering with liveliness as a physical status*).

(By the way, l*ogically, there should be many undetectable momentary liveliness of micro-state livings in many life-supporting natures.*)

Even though, at this moment concerning improvements, it assumes that the bio-living species are the fast responders from the micro-molecular level, realizable when measurement performs on comparable strategy from the appearance or birth, growth, and exhaust or disappear within the period of existence as physical status like a human being.

{(Indeed, one cannot deny that the living molecules are active since they evolved from life-orienting energy and can communicate for ***survival necessities*** *in a neighboring friendly atmosphere.*

Or the same, from the viewpoint of energy, the sense of ***survival necessity*** *can be defined as the struggle to reserve the zero-status-stableness of the mass-energy frame/s (mc^<2>) for distinguishing bodily appearance molecules.)}*

Accordingly, since the inception of molecules, the changes in the physical constitution at the energy levels initiate following different forms of alternation to many realizable and unrealizable courses. Then progressively, their unifying appearance as primordial humans, and then to reach these modern intelligent human beings with **advanced communication techniques**, using natural, non-natural, and the combination of both that accepted as the sense of scientific or knowledge-based evolution by the contemporary bio-intelligent generations.

h1. Clarification: -

The modern **web in networking communication** is one of the **advanced** technological samples invented, developed, and categorized by the specialized efforts of intelligent beings following the best planning, analysis, systematization, designing of wireframes, coding of software, etc., before speedy delivering through networking means from nearer to distant web communicators that allowed to maintain the reliable and close entangling link regularly, which is entirely unlike the olden days physically attending communication and then follow the proverb like 'out of sight, out of mind' of entangler and entangling relations with something.

However, during the olden eras, the above modern generation's wire network and remote networking communication systems were impossible to believe, which is somewhat reasonable to apply as an example of entanglement with the far-distant relations of most microparticles at their state of appearance.

i. Indeed, one should not forget the gift of our prehistoric predecessors since they had acquired many items besides communication techniques following natural practices and non-natural creations for survival necessities in those fierce ancient environments.

Those challenging eras' experiences carried forward through celestial or energy levels, which influenced and allowed for augmentation stepwise since the energy levels biologically positioned human species are somehow entangled from generation to generation regardless of their appearance justified as ancient or past, modern or present, and advanced or future eras generation. That somehow resembles the contemporary invention of biological **genetic linkage** (*that is, the closeness of genes*), which means carrying forward genetic information, even one's living state and physical appearance exhaust but bodily pass to present one and follow on to generations.

This natural technique for comfort communication or transporting genetic information by the biologically positioned species may relate to quantum-level tunneling effects since it follows the **genetic linkage** method from one generation's body mass to other following generations, tirelessly years after years of biological existence.

→ *{(One thing is to remember that the word 'miracle' is only a satisfactory application to living physical systems.*

But in energy levels, everything is possible, whether someone can analyze or not through the integrated brain processor for realization as practical or logical for acceptance, or consider as fantasy or fallacious.

Undoubtedly, the imaginative faculty or mind's eye of human beings is wide-ranging from attainable to impractical, based on the specific way formed physique and developed consciousness that solves some purposes while existing in any supportive surrounding.

*But it does not mean someone or something (*that is, living or non-living state) *cannot solve one's impractical imagination.*

Deliberating the purpose of a brain processor, which natural ways set in the physical system mainly solve and secure the requirements of molecular unions portrayed as lively body-mass so the delimited consciousness can grasp or attain comfortably that configured in a brain process.

But the rest of the impractical imagination yielded due to molecular-level activities in the living being, standardly declared as impossible and weird fantasy by the brain processor since the limitation of preset consciousness or acting mindset.

However, reflecting as a living body-mass, from the standpoint of energy and concerning the sense of flexibility, the pre-stage mass-energy frame is more flexible than the post-stage's energy-mass that transmogrified into a lively physique, following the proper abidance of naturally derived atmospheric physics in a fitting habitable zone/s.)}

j. Whether someone accepts or not, from the standpoint of energy, every individual of physical existence is integrally regulated due to balancing applications of internal and external energy physics, whether resides in a native atmosphere or migrates to a non-native inhabitable atmosphere.

Likewise, the word senses and applications of information, communication, entanglement, etc., is not modern-day discovery. But in lieu, all deeply interlinked with energy and its diversifying roles from nearer to an unimaginable distance like nowhere feelings.

ix. The drawing EU64 is an altogether the graphs EU04 and E30 defined in the earlier book (Part 3).

That indicates near and plausible distant entanglement between some limited microparticle proprietorial specificities, which are presented to display as one kind of linguistic elements /alphabets (*in the drawing, conveniently used as English*) with different typefaces, assuming their individuality and distinct characteristics in one block.

This specific type (*English*) of the linguistic element in a block of mass energy differs from the many other contemporary and obsolete languages (*some kinds of linguistic scripts mentioned in the graph EU62*) and linguistic alphabets situated near and distant rectangle shapes or blocks of mass-energy as shown in the same drawing EU62.

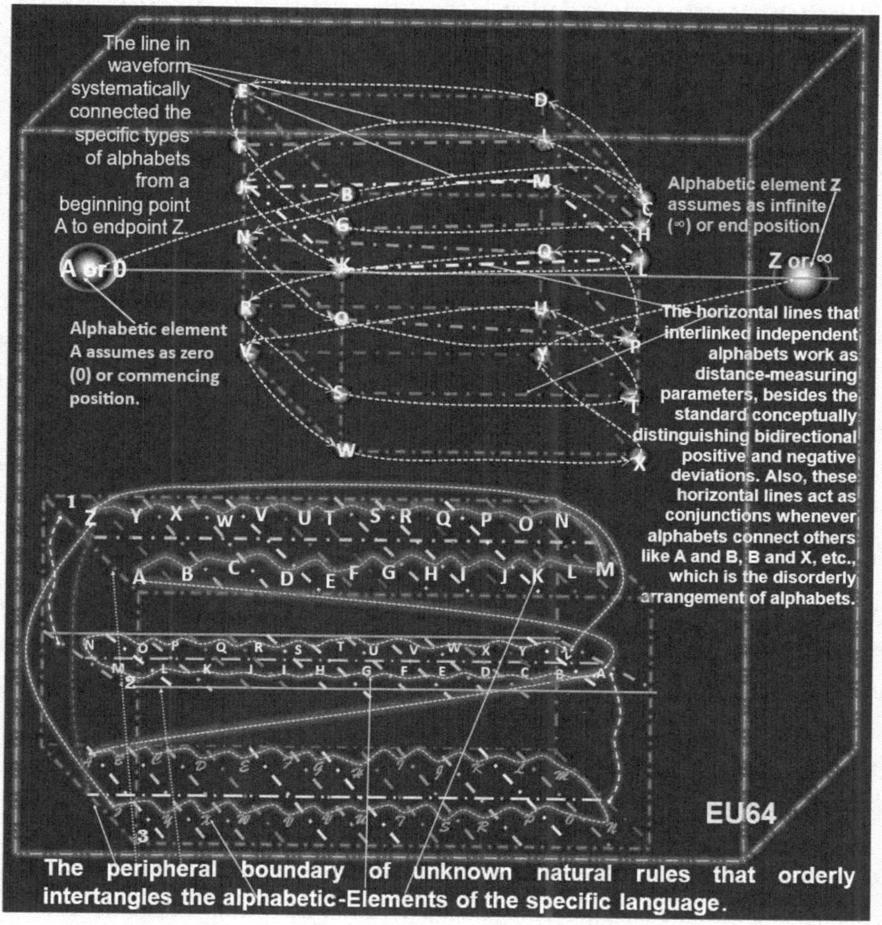

Also, presume that all the blocks are filled with more or less coequal classes of energies so it resolves a common purpose and application for proprietorially distinct microparticles (or *alphabets*) sited in the fathomless 'energy- Ocean' of the 2nd-energy-belt.

a. Looking at it from an energy standpoint, assume that technically skilled (*human*) web planners (*about web planners, already mentioned in the earlier point*) serve as energy physics in the rectangle areas or blocks (*either EU4 or EU30*) covered by a particular class of energy that momentarily takes care of and administers the residing microparticles following natural rules of energy Physics containing only one type (*that is, limiting by parametric value*) of energy.

→For illustration, the most microparticle levels practices of transportation of information justified somewhat similar to the orderly positioned near and distantly entangled alphabetic elements.

But for their disorderly placement (*shown in the drawing EU61*), like for making valid words, meaningful sentences, and other applications in the coded form, human web planners have to properly organize beforehand for relaying as packages through networking means to many types of linguistics subscribers those who entwine with a particular web domain.

It means a typical group of microparticles in a place of locale or rectangle block may pass information directly, as well as indirectly, to distant blocks of related groups through the faster energy networking media, provided energy physics of a native block could organize to carry through the support of external energy physics.

→Indeed, transportation of datagrams needs to follow non-natural or technological means, but for the case of microparticles like E0's or having the lower to the mass value's unit/s able to flow through the faster energy networkings, provided natural physics of a locale or in rectangle block can arrange appropriately for transportation through external atmospheric physics.

→The web planner can call for changes in the endpoint of users' practices since he got the technical right, which is easy to pursue directly and justify through living sense as natural instruction.

But it is somewhat like the energy physics of a **locale atmosphere** (*i.e., the atmosphere derived by the amount of energy occupied in the rectangle block considered as* ***locale***

atmosphere) that can call for changes to the containing mass form energy, microparticle articles, objects, etc., in coordination with surroundings atmospheres.

(Technically, one thing to quote is that any form of premature mass-energy frame analogously has more advantages since, at the elementary level of energy, mass energy is more flexible than structurally depictable microparticles.)

→Further, the above web planning networking communication is unlike the case of distant ancient or obsolete languages from contemporary recognition. But for some reason, many people in different countries still prefer to use those virtually outdated languages. Thus, those **obsolete** items need to follow somewhat indirect, besides direct sharing methods of communication about the changes or alteration by the web planner or energy physics of a locale.

But for the case of energy products like microparticles, nothing can justify as **obsolete** since every form of microparticles, particles, or objects appears and disappears in a recycling manner to a near locale or distant isolated atmosphere.

(One thing to remember is that energy and its progenies or energy products stay like differentially distinguishable isolated worlds since the Energies have the highest level of personality in all aspects than the object forms progenies and sub-progenies.)

x. Further, for the instances with different distant linguistic alphabetic elements that entangled for expression purposes of the human species, and to match with the most microparticles for proper and faster communication roles. I want to recount the energy levels **influencing energy elements**, which are the closest degree of thinner compared to the motion element in any mass-energy frame regardless of its (*mass-energy frame*) shapes and sizes.

→In this regard, conveniently may remember the elemental arrangement in a mass-energy frame with the drawing EU65, *which is illustrated in earlier books, and about* the **influencing energy elements** that perform the 2nd positioning roles on the aspects of speed and neutrality from the motion element and tend to sway in the fastest manner directing for immediate response to other energy elements in the frame of mass-energy or an object.

Likewise, the motion element and influencing energy element of a mass-energy frame, the ostensive Purest-energy in the 2nd-energy-belt works for systematizing roles of fathomless energy-Ocean that all fall in the formulaic frame E = mc^2 or augmented E = mc^<2>.

(By the way, applying a philosophical sense to the equation E = mc^2, it may quote that in any mass-energy frame (E), the diversifying features of mass (m) regulated by the constancy of speed or motion (c), whatsoever the volumetric area cover following the essence ^2.)

a. As indicated in the drawing EU62, different linguistic elements or microparticles formed for one standard purpose like communication, which took position in various rectangular shapes and followed the natural roles of their local energy physics activities.

Since a microparticle (*or assumed linguistic element*) is analogously lighter than its associated massive mass-positioned objects (*or words*) where internal energy physics is consistently busy to tackle the individuality as an object facing more resistance than a microparticle *(similar to the words and unitedly arranged sentences function as a heavier text parcel than characteristically individual alphabet or element).*

→Also, it may be mentioned that in a mass-energy frame EU65, the influencing energy element is the 2nd in order of thinner degree (*motion element is the 1st in order*) than the rest of the energy element and has influencing and faster roles than all other different roles of energy elements in a mass-energy frame.

This drawing is taken from one of the earlier books to indicate the minimum way concerning the arrangement of multiple energy elements in a mass -energy frame representing an object.

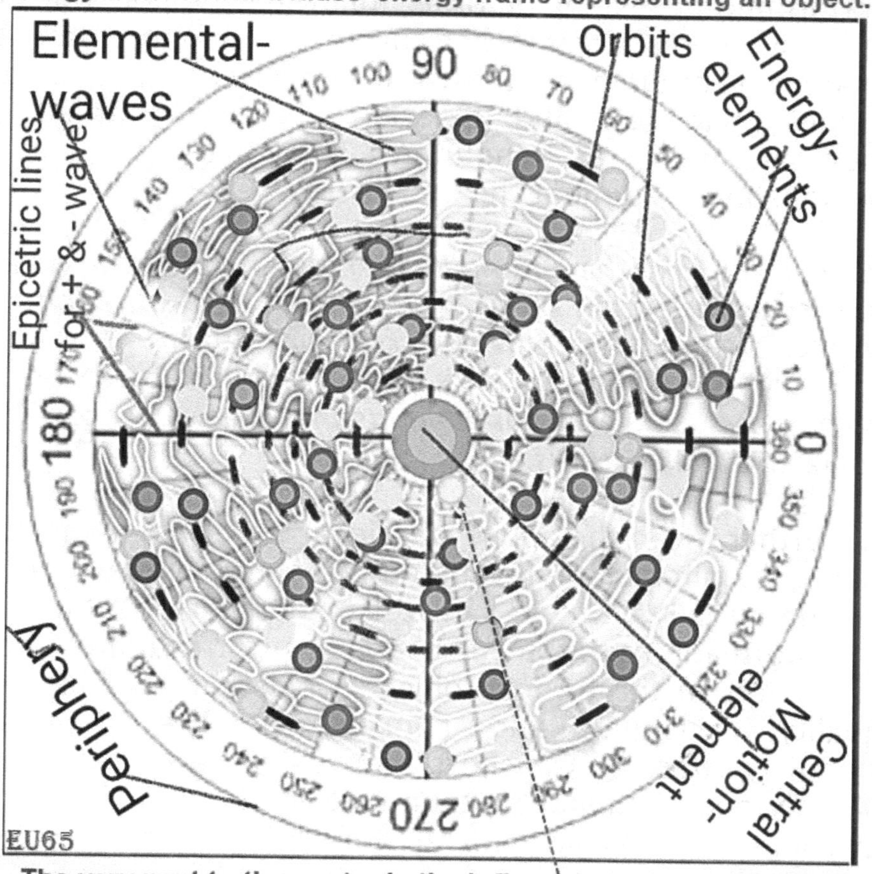

The very next to the center is the Influencing energy element.

So, based on the above analysis, briefly say, any energy signal (*assumed as waveforms*) generated by the microparticles can follow a quicker energy levels' networking path since, compared to the overall constitutional roles of a microparticle (*that is, a mass-*

energy frame), its released energy signals will be a fast and efficient traveler.

b. Further, the word sense **tunneling** applies mainly to the case of low-mass particles and may also justify its use for the matured tiny or bulk state mass-energy frames.

But, merely at energy levels, the word sense **tunneling** seems inappropriate since energies are everywhere, performing in systematic order following the holy roles of ostensive Purest-Energy that covered nook and cranny as imaginable or unimaginable most-micro levels mass-energy frames, particles, and objects, whether regarded as momentariness or enduring individual long-lasting personality throughout the 2nd-energy-belt.

{*Something, like the entangling personality of the mass (m) and speed-orienting motion element (c) as a mass-energy frame, maintaining equivalent relation with the 'E' (that is, E = mc^2), whatever and wherever their discernment recognizes as fleeting or long-lasting of appearance with some structural status.*}

c. Further, from the above clarification, at this moment, it seems a minimal representation comparing the modern technologically developed miscellaneous way of entangling the information and transportation by using multiple media or carriers in the form of objects or datagrams following quality packaging and management like shipment, shipping, transit, transport, conveying, and freightage directly and prominently countable methods or indirectly invisible electronically generated networking links.

d. Also, from the standpoint of energy, it assumes that a microparticle is the primary stage of a mature unit-value mass-energy frame, which gradually organizes due to proper effort of internal and external energy physics, which in turn results in a physical structure before journeys toward dimensionally discernible particles and large objects.

And accordingly, at the energy levels and for the primary stage of a mature unit-value mass-energy frame or the early stages' unmatured microparticles, the ostensive Purest energy will be the best and instantaneous, on the point of entanglement and method of quickest transportation.

Example: -
Suppose one of the twin-born Time travelers covered to reach millions of light years while his physical counterpart touched old age on this Globe.

Here, the constitutional physical times (which may justify as 'energy-Time' of living biological physique) of the Time-traveller twin comparably slowed down to the physically deriving times of the counter twin, who is staying in this bio-Globe.

Also, many intelligent beings are aware that the faster someone travels, the slower the experiencing time (that falls in line while on physical analysis). But it means time flows depend on the rapid physical momentum.

(Indeed, many energy levels' physics values criteria are unaccounted for when applied to that physical analysis, so many stopgaps average out for the satisfaction of human perception.)

→Further, believe it or not, following individual in-built biological clock time, which may justify as a regulator for one's physical systems, is in concurrence with the surrounding atmosphere when performing as a lively human being.

Besides, in the physique or the body -Clock, the integrated molecular level's communication techniques by releasing energy Signals are comparably faster methods that allow one to maintain a perfect union with a healthy physique in a conducive nature, comparing the brain processor that mainly dominated for the time-based (*here time-based means, constitutionally set biologically deriving energy-Times*) measurement of energy-Signals derive by the delimiting sensing faculties.

Chapter 17

About objectively defining the word-sense Singularity and the Singular, and its clarity for appropriate application to energy in the 2nd-energy-belt.

From the dictionary interpretation of the word, the singularity illustrates the state, fact, quality, or condition of being Singular. But without decoration, the same sense signifies that everything exists in one theme /configuration.

i. Approachable /observable singularity.

From the opinion of classical physics and mathematics, it is a point at which a function takes an infinite value. There are many places in the observable Universe where the laws of classical physics straightly break down.

In general relativity, coordinate singularity works in one coordinate but disappears for True Singularity.

→The primary purpose is to present the word singularity and how objects depend on parameters when properties, especially tackling small-level variations.

→Also, there are many interpretations of the word singularity, following the perceptible items in the 2nd-energy-belt. But most of those interpretations find out and indicate the physical impossibility of reaching a point of singularity.

ii. Inaccessible /imperceptible singularity.

For the easiness of our understanding, it recalls that the property of energy is invisible, colourless, and shapeless unless it governs in the form of the mass-energy frame (mc^2). Also, in common, it recognizes through the sense of indestructibility.

→And by no means 'no quantity of energy' can be produced in excess or borrowed over and above that occupied /settled in the 2nd-energy-belt upon maturity.

→From the viewpoint of energy, singularity considers where all the energy -components or energy elements are reserved as a unified zero-status-Absoluteness position, or say, Ture Pure-energy as the infinite status of presumptive sense or as a property of **True 0^abs** (*non-observable sense*) in the Supreme - Universe.

(Note: Absolute-zero, or '0abs': -
The intelligent figure Absolute-zero, dignified as '0abs' practically stands at an absolute ceasing point of all ideologies.

Another way this may say the living conceptions obtained an absolute base point to the satisfaction for easy perception on comparable manners. Because, at this stage, all the ideological comprehension is entirely exhausted or diminished at timelessness status.

Now, if this (0abs) augments by a circumflex sign (^) to indicate an eternally immeasurable standing. Then its importance may justify as ***'True 0^abs'*** *and possible to implement for the Pure energy of the Supreme - Universe.)*

→Or say, in the Supreme - Universe, all workable and non-workable properties of energies include derivable differences relaxed at a 'state-of-Absoluteness.

→The righteousness of True singularity gradually interrupted upon initiation of the Universe. But probingly, it retains as the gift to the alive bio-intelligent beings while entertaining the observable world upon maturity of the Universe.

iii. So, conveniently, the existing equation E = mc^2 seems very appropriate to apply for defining the initiation of the Universe from the True singularity source (*that is,* ***True 0^abs)*** where 'E' represents the eternal status of the Supreme - Universe. And 'mc^2' symbolizes its breakaway belongings diverse mass form energies and the very primary stages frames of mass-energy that contributed to the preparation of the Universe.

iv. The 3rd-energy-belt is a True Singular position in the matured Universe.

In the observable Universe, expression through the word-sense singular, in common, distinguished by superiority as one or single of a kind. This type of expression through the word singular position is a satisfactory conclusion for object and objectual products to the sense of living intelligence.

→But from the viewpoint of energy in this Universe, only the 3rd-energy-belt is an absolute-Singular position contained with the extreme-finest form of Purest-energies and solitary status having no possibility to segregate or distinguish on the point of differential value. This Singular position is unconditional in all imaginable or unimaginable aspects.

Chapter 18

Significance of 0 /Zero and its purpose of the application for observable Universe as well as Energy-Universe in the matured 2nd-energy-belt.

In general, expression through this numerical symbol 0 (*'zero' functionally represents a whole number*) indicates the sense of nothingness or absence of anything on the point of size and quantity.

But, for all mathematical computations, 0 works as a primary factor and an acting origin for diversifying extension of natural (*natural numbers have positive integers and include numbers from 1 till infinity*) and non-natural (*non-natural numbers carry negative, fractions, and decimal values*) directional degree as right-left, left-right, up-down, down-up and so on extendable to the acceptable on logically perceptible sense by the bio-intelligent living species.

i. 0 is the first whole numeric value that brought a change of perception to the intellect beings. Also, its application to a grassroot level became essential for all concerned in our intelligent society, like math, science, economics, politics, etc.

→Indeed, in the historical eras, the 'zero' is represented differently as a placeholder.

In the 7th Century, the great Indian mathematician and astronomer Brahmagupta regraded to employ zero as a number with its independent rights.

ii. From the notional and mathematical viewpoint, any independent integer 0 /zero has exclusive disposition from the point of arrangement on the sequential gradation of other whole numbers from 1 to 9.

→0 /zero resolute as no value /neutral when placed before a number. So, it can consider the base for all other integers from 1 to 9.

Besides that, unlike the other numbers' flexible attitude upon application of mathematical symbols, in general, '0' does not affect at all since it acts as a constant value or position.

iii. So, harmonizing the specificity of bio-intellect species, ideologically, the essence of 0 /zero seems a comfort to convince and relatively convenient to apply beyond microparticle physics to energy regardless of the mass-energy frame and its constituent energy elements.

→And accordingly, in the earlier two volumes (*Mythology, Philosophy, and Science, Book for the New Generation*), this 0 is befittingly applied for the works of energy instead of discovering new symbols, numbers, or uncommon word applications.

iv. Further, the above may extend with a graphic representation (EU66) concerning the number-integer /number-element 0, which is conveniently considered a baseline for most bidirectional portrayals.

→On multidirectional depiction, this "0" may assume a 0-positioning base-point instead of a 0-positioning baseline.

Since a simple bidirectional sketch "baseline" commonly bifurcates ideology into upper and lower value directions.

But for multi-direction, the base point 0 /zero acts as the solo origin for multiplex variations for all physics-value of any unit-shaped mass energy and even the Elemental-value of its integrated components /elements.

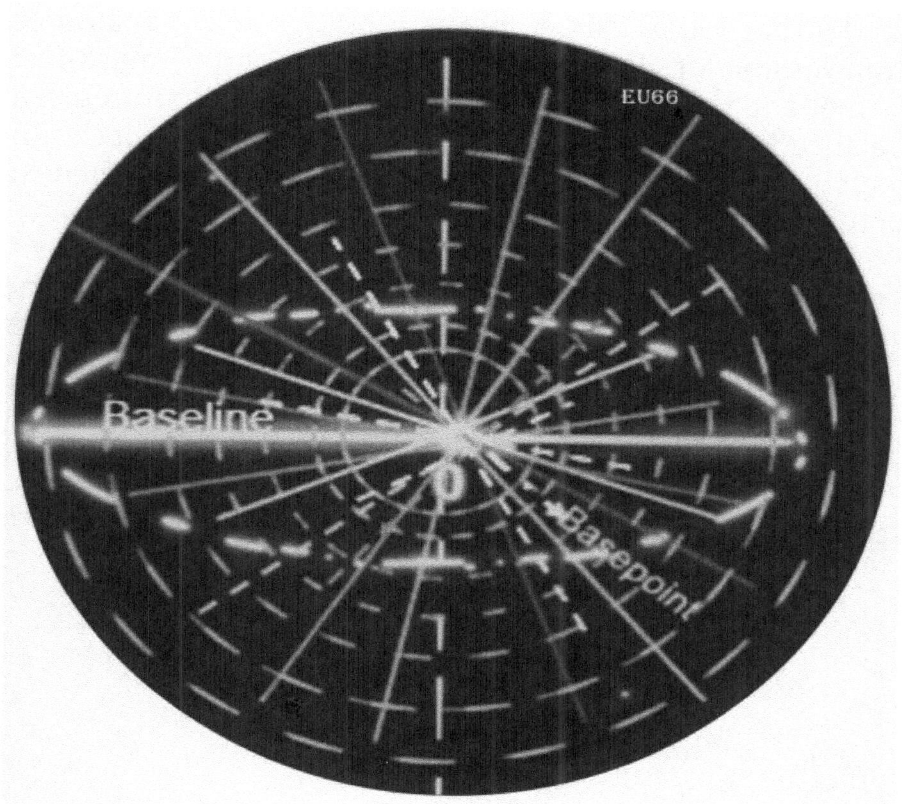

→Also, conceptually, the word-sense 'line' usually implies some extending value compared to the word-sense **point**, i.e., the '**point**' is a solo (*or singular state*) identity or having no imaginary alternative for any diversity, whatsoever the recognition of the line (*that is, add on with multiple dot points to make a line*) casting with specific words like curved, parallel, jig-jog line, etc.

v. Graph EU67 is for easing comprehension about an admixed unit-value energy (mc^2) with complexly wired energy elements that function multi-directionally, justifying the significance of 0 or centrally located motion deriving motion element as a base point.

a. Further, the same graph (EU67) may analog to commission or undertaking unit-form energy, mass-form frame, and their unitedly positioned product/s strategically balanced with the aggregated amount of energy, under the flagship or standardness of zero-status-stableness, predisposing-constancy, and likewise conjectures.

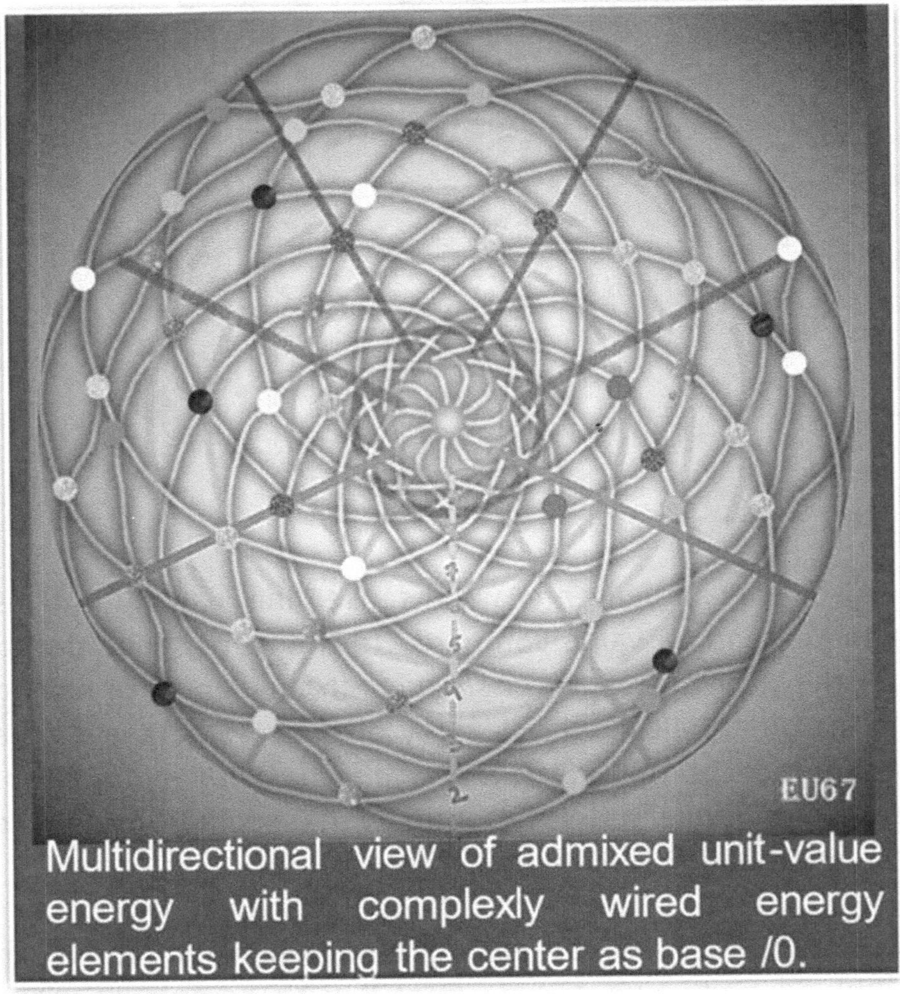

Multidirectional view of admixed unit-value energy with complexly wired energy elements keeping the center as base /0.

→Although the drawing followed the foundation level of bidirectional ideology, even the same may justify as deemed fit to link with multidimensional requirements for reckoning the multi-pronged strategies to three-dimensionally specialized intelligent beings.

→Besides, it is a reasonable position to understand the working relationship of constitutionally developed energy elements and their eternal versatile activities in a conditionally prescribed unit-value energy or a prospective object.

b. Concerning the versatile activities of energy elements in a mass-energy frame, it may exemplify the significant positions and proprietorially diverse status of "**liquid**" in this nature.

Water is usually recognized as a state of matter and determined by volume but changes its constitution into various recognizing levels like solid, **liquid**, and gaseous with distinctive effects.

→However, following the energy, it may say, particularly dignified liquid, is complexly shaped from specific types of energy akin to water, acid, oil, and so on of their strategic structure as liquid form. Yet, their aggregative energy values are distinctive from each other and one another, therefore the distinguishing proprietorial display.
And accordingly, the independent features and even distinctiveness from the point of design or configuration of integrant energy elements.

Relatively, the same may represent as an example with the classically recognizing various types of energies following the observable properties like chemical, nuclear, thermal, gravitational, electric, and so on with the suffix energy.

vi. In addition to the above, 0 /zero can logically act as the physical singularity but entirely unlikeness when reflected on the point of energy.

From the definition of the word sense singularity, the common expression represents the condition of being singular.

Scientifically, expression through the word 'singularity' is realized as the function taking to an infinite value or point where some property falls at an infinite / ∞ /unknown position.

All those ideas somewhat classically represent the significance of Zero in a compressed positioning for the origin of all diversity of states of matter. Whatsoever it consciously expresses with mathematical symbols keeping two constant positions, one side as zero/0 and the other is infinite/∞ for ideologically representing expansion and contraction due to explosion and implosion of values /properties of something, obeying the self-bearing constancy as an individual being.

a. As a bio-intelligent species, our experienced consciousness regulates the limitation of direct nine mono-state number elements *(from 1 to 9*), keeping the symbolical '0' a constant for all diversifying activities. (*Indeed, for binary eras, only two numeric elements like 0 and 1 work as base factors for all diversifying placements. Something like '1' is true or "on, and '0' is false or "off" for logic computing.*)

Even beyond the delimited mono-state nine number figures, consciously and intelligently manipulates mono status elements (*including 0*) with left and right combinations to depict larger denominations unto reach a consolidated position both actual and potential that might seem infinite.

→But for all those cases, the self-bearing constancy always obeys as an unaltering opinion regardless of individual position or the species as living beings, which evolved following the genetical acquirements and natural ways gained experiences while existing in conducive atmospheres.

a1. The drawing EU68 shows an example combined with figures 1 and 9 to represent cardinal values such as nineteen and ninety-one.

Logically, values 19 and 91 are in the compressed state of 1 and 9, and numeric values 'one' and 'nine' alternately act as ascendant or positive and descendant or negative in order of individuality.

From the standpoint of energy, it also means the abridgment of many functional probabilities between the numeric ranges '1' and '9' and their combining outcome to match for realization through the three-dimensionally expert physical position of bio-intelligent beings.

→It will be easier to understand if someone can imagine a graph (EU68) of concentric circles with nine orbital positions following the degree of sequences from 1 to 9 and then beyond. Then, the combination of all probable number elements can decompress (*or extend by using positive essence*) to wide ranges to an unknown /infinite, and in reverse manner possible to compress or narrow down (*by using negative sign*) to the most micro position or '0' to understand the large variable computations while concealing /condensing many fictional properties of True energy levels physics.

a2. Further, based on numeric elements and by directionally discernible negative and positive assignment: -

►From a directional viewpoint, integer +1 is closer to +2 but extremely distant to +9 and stepwise to infinite.

Some Palindromic number figures remain identical even though they are placed in a reversed manner to understand their overall increased values from an individual. But, from the viewpoint of energy elements, I may define the duplication of several number elements to comprehend the overall status encapsulated by large values. Those are like the 11, 22, 33, 44......& so on. One thing to quote from the earlier volumes, in the purview of momentary property of energy elements, cannot duplicate unless or otherwise advantageously manipulated for the convenience of an intelligent system, keeping the mystery persist in an abridged state

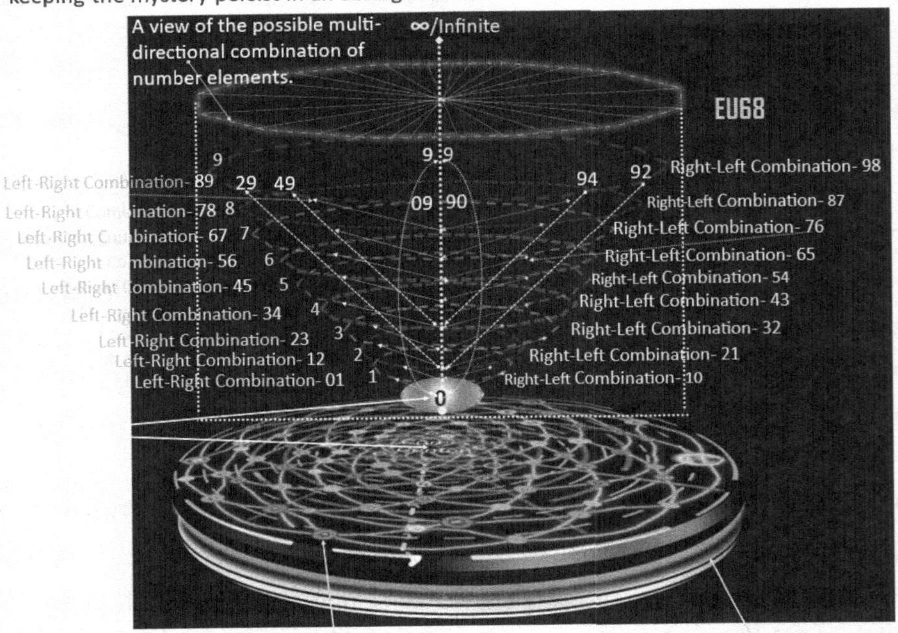

Following the 3D projection, this drawing is a partial view to indicate the versatile multi-directional roles of energy elements in a compact mass-energy frame.

►Likewise, the negatively bearing number integers /elements -1 is closer to -2 but furthest outlying to -9, and so on negative discernment up to an unreachable infinite (∞) value or position.

►From the base "0", -1 is the nearest lower (*or negative*) to the -9, and +1 upper /positive to the furthest +9.

►The above conception for integer/s can extend from the decimal viewpoint {*or may mean dilate in descending order through decimally accountable numeric elements by using (+) and (-) values assignments*} for ease of computation and understanding.

→Also, all those individual numbers and number symbols customarily or assumed as rules manipulated by bringing their relation very close to far for ease of computation and satisfaction for delimiting 3D-based consciousness.

→Likewise, dimensionally depicting Uno, Duo, Trio, Quattro, Penta, etc., of appearances, one and the other way around, all integers may signify with diverse acts and effects despicable as 'pro and anti-values' activities to the consciousness of an individual bio-intelligent. But all those nailed as term-bound following regulations provided through three-dimensional natural implications in a measurable system.

b. From graphs EU68 and EU69, it may be easy to grasp how the singularity works to the consciousness of classically presented observable items that seem far beyond contemporary logic sequences when considering the unobservable energy level and their operating energy components. Also, some Palindromic number figures remain identical even though they are placed in a reversed manner to understand their overall increased values from an individual.

→But, from the viewpoint of energy elements, I may define it as the duplication of several number elements to comprehend the overall status encapsulated by large values. Those are like the 11, 22, 33, 44……& so on.

c. One thing to requote from the earlier volumes, in the purview of momentary property of energy elements, at the energy levels, nothing can be duplicated unless or otherwise advantageously manipulated for the convenience of an intelligent system, keeping the mystery remain an abridged state.

Like a compact state unit-value mass-energy frame acting as an object of solid status. In this case, for the object/s, an overall outcome of physics is due to the integrated act of energy elements, irrespective of any standardness of parametric values' delimitation or beyond.

d. Something analogous to a unit-value object or matter constitutionally composed of tiny elementary particles with distinctive properties organized in the periodic table. Any unit-value object /matter must occupy a mass of energy in a compact state that gives a dimensionally projecting frame. So, regardless of whether orderly or disorderly the periodic table of elementary particles is, its combining properties will be the cardinal property of the unit-value object /matter.

e. So, in the depth of computation and findings of objects and easily discernible items, mathematically representing 0 can act as a singularity, center point, or constancy as mainsprings for all diversifying rolls as rise, fall, up, down, left, right, positive and negative of expression based on the properties and application for mathematical essences.

→Also, following the characteristics of the observable Universe, there are many interpretations through the word-sense singularity. But, most of those interpretations find some inaccessible or unanalysable point/s indicating the physical impossibility of reaching the True energy levels' singularity.

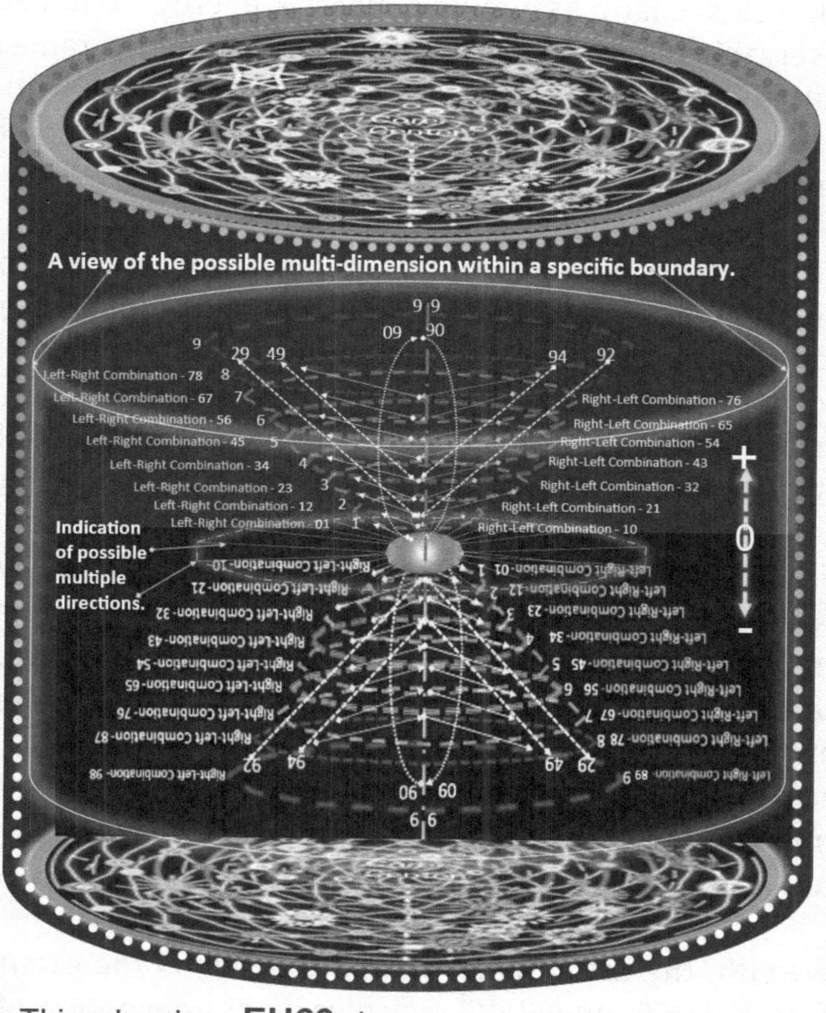

This drawing **EU69** is an extended position of EU68 projected following the 3D purview of a minimum possible complex scenario assuming massive energy -cloud representing a mass-energy frame and multi-dimensional activities of its constituent working essences or energy elements, besides multiple objects playing within the energy cloud.

EU69

f. The drawing EU69 is an extended position of EU68 projected following the 3D purview of a minimum possible complex scenario assuming a massive energy cloud like an observable Universe, representing a mass-energy frame and multi-dimensional activities of its constituent working essences or energy elements, besides multiple objects playing within the energy cloud.

→The drawing EU69 also indicates that the numeric base computation system is very convenient and comfortable for judgment based on the three-dimensional established brain processor, which mainly manipulates and yields in the curtailing manner of many near and distant physics values activities just for the satisfaction of bio-intelligent beings.

vii. For universal purposes, three-dimensional ways invented Zero /0 and Infinite /∞ symbols are reasonable if someone intends to apply the sense zero and infinite the other way round, which means the ideational sign 'infinite' is possible to implicate in place of 'zero' securing their positional logic to overcome the dimensionally materialized physical fantasies to understand the fundamental works of Energy-God.

Herewith, the drawing EU70 (*this drawing is the extension of EU68, EU69, and EU40*) may give a possible notion of the application of 0, ∞, and their parametric augmentation securing their 3D positional sense, unchanged.

a. EU70 is a Simple Projection and Surface View of an Observable Universe, assumed following the fundamental equation E = mc^2, where if E justifies as an observable Universe, the acting (*not True*) 0^abs the containing m (mass) and c (*motion deriving energy or comparable most-thinner in the observable Universe*). In the observable Universe (**E**$_{\text{observable-Universe}}$), the most-thinner form of energy (c) acts as an unchangeable constant or like 0 for the entire amount of mass occupied by E.

a1. Herewith philosophically, for the overall coverage of the observable Universe, '$E_{\text{observable-Universe}}$' act as infinite for its holdings (mc) regardless of representation as discernible, non-discernible objects and observers like human, alien, those who are pre-set based on particular dimensions (^<2>).

→But, externally of **E**$_{\text{observable-Universe}}$, for any fast-moving object like an observable, non-observable Universe or interstellar traveler, '$E_{\text{observable-Universe}}$' may act as a 0, regardless of the quality and quantity of m and c or their encapsulated diversifying objects.

b. Healthy objects or matured mass-energy frames (as shown in the EU70, E_V90=mc^2, E_V80=mc^2, E_V70=mc^2, and likewise) are systematically maneuvering due to regulating abettance of the most-thinner form of energy or the acting 'c' that covers throughout the 'E' observable-Universe.

{(*In brief, the word-sense* ***healthiness*** *of an object or mass-energy frame determines its proper administration from the internal elementary levels of deriving energy physics, in concurrence with the surrounding atmosphere to secure the balancedness of individual positions.*

So, the unhealthy situation of an object or mass-energy frame can fluctuate irregularly within the gravitationally controlled ***orbital*** *range* **(orbits**, in general, administrated by the reasonably thinner form of energy or 'c' when some amount of energy secures its position as zero-status-stableness in the surrounding atmospheres whether it resides in an observable or non-observable Universe in the energy-Ocean of the 2nd-energy-belt**)** *before it collapses and transforms to some other status*.)}

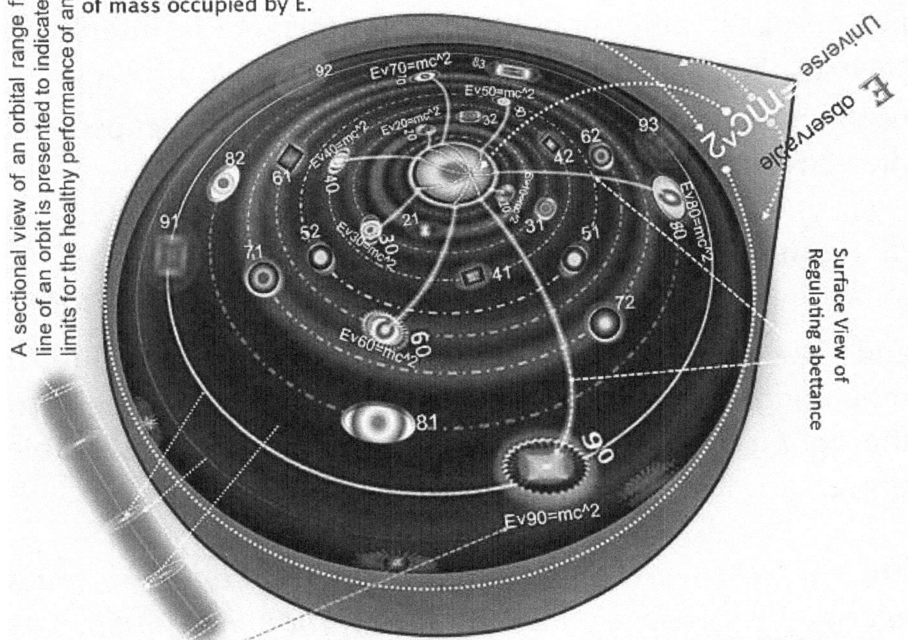

EU70 is a Simple Projection and Surface View of an Observable Universe, assumed following the fundamental equation E = mc^2, where if E justifies as an observable Universe, the acting (*not True*) 0^abs the containing m (mass) and c (*motion deriving energy or comparable most-thinner in the observable Universe*). In the observable Universe ($E_{observable\text{-}Universe}$), the most-thinner form of energy (c) acts as an unchangeable constant or like 0 for the entire amount of mass occupied by E.

Healthy objects or matured mass-energy frames (as shown in the EU70, Ev90=mc^2, Ev80=mc^2, Ev70=mc^2, and likewise) are systematically maneuvering due to regulating abettance of the most-thinner form of energy or the acting 'c' that covers throughout the 'E' observable-Universe.

viii. Also, dilating the equivalent (=) relationship of 'E' and 'mc^2, where assuming E acts as '0abs, then both the E alphabet and 0abs value conceptually will perform as an absolute constant position for any diversifying attitude of m (*mass*) and c (*momentum*) within any simple or complex dimensional status of the united supremacy of (*mc*).

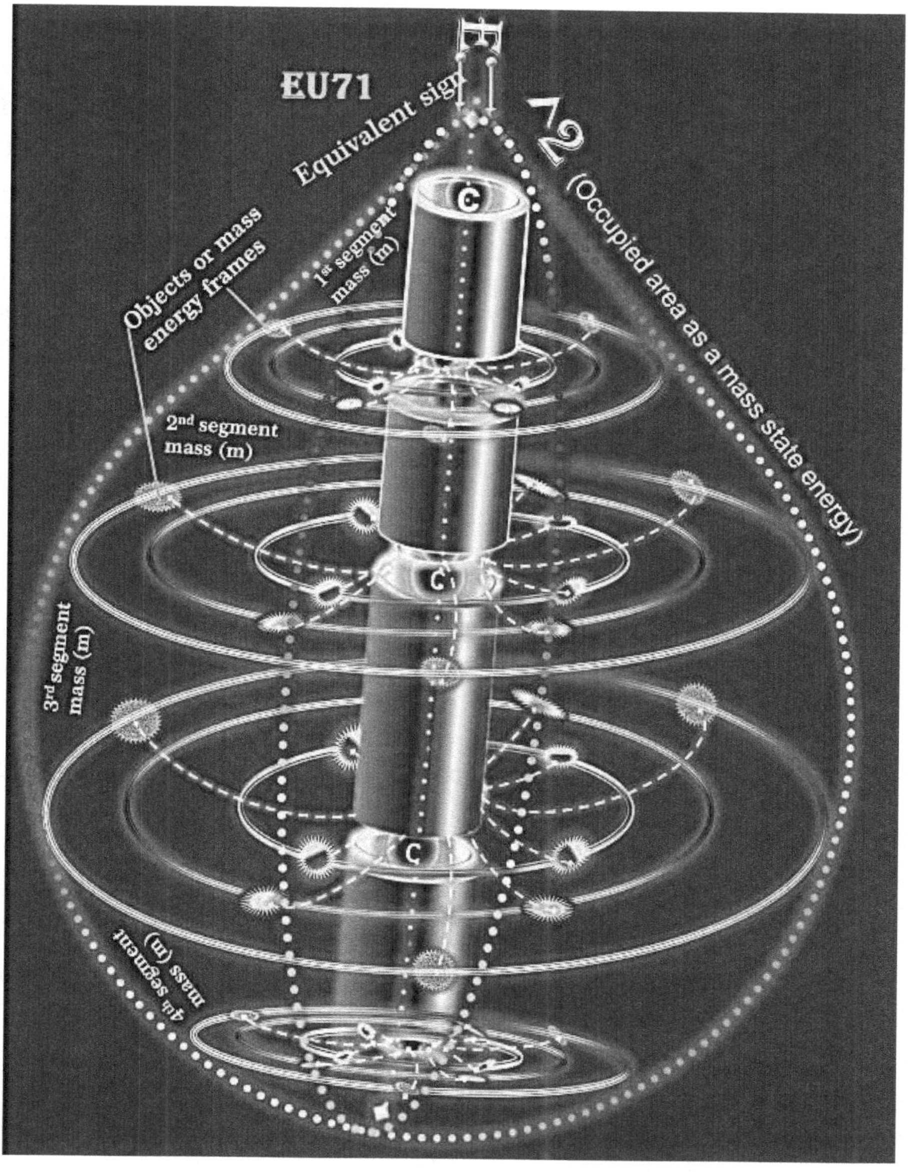

The drawing EU71 indicates the unchanging equivalent relationship of bonafide state energy 'E' whatsoever 'mc'

rein in with small or massive amounts of energy and represent somewhat like premature mass-energy and post stage's clusters mass-energy frame those are ready to turn into objects or already evolved as objects.

In the drawing, the equivalent relation indicated by the two parallel arrows represents as = (*equal sign*) may think of independent directional aspects to make a balanced relationship whenever and wherever mass-energy depicts as a referenceable frame/s either of a '0' dimension or the constant position like 'E' and then widen its relation by elevating steps (*in the drawing steps are indicated into four segments*) in a directional manner like the conceptionally 1D + 0D + 1D =3D standard for modern human intelligence that again may extend multi-directionally following the diversifying scenarios derived due to the relative relation of mass (m) and motion (c).

Chapter 19

Recollection Of Some Points on The Aspects of The Construction of The Universe.

i. Three barriers (drawing EU72a) of the 1st-energy-belt.

By the way, concisely remember some points about the three barriers of the 1st-energy-belt that taken a beholden position to secure the original dignity of the Universe at a healthful buoyant state within the Supreme - Universe. Even though the multi-directionally regulating physics values diversity, including the most significant application of Gravity, determines the stability of the 2nd-energy-belt whatsoever accountability for construction, destruction, expansion, contraction, and so on following the structurally recognizing some particular dimensional (*1D, 2D, 3D, 4D..., etc., framework*) ways of measuring methods reserving one or more perceptibly determining or inferring variable constant like ‘0D’ at a fixed.

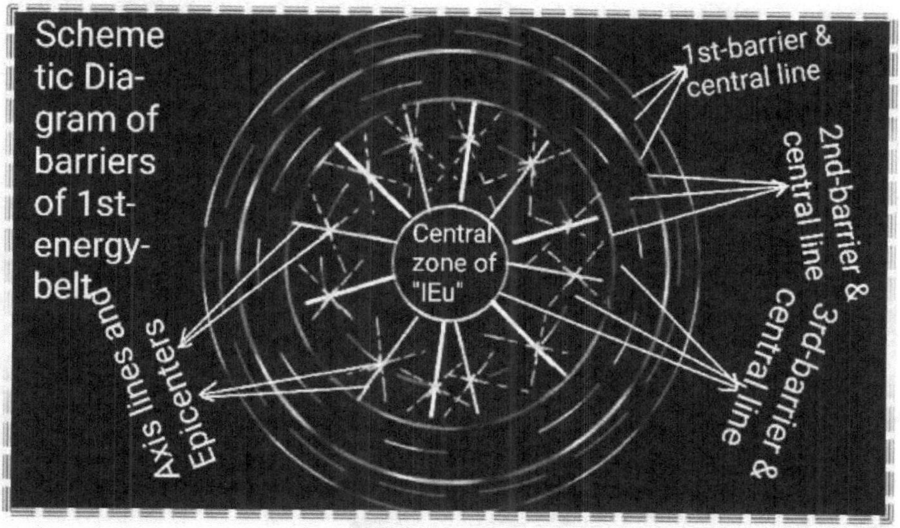

EU72a

Further from the earlier volumes of the book "Mythology, Philosophy and Science" regarding the three barriers of the 1st-energy-belt, here are some points to note with drawings (*below are the collections from the earlier book, volumes 1 and 2*).

a. While in the progression of the immature Universe, some comparably thicker energies gradually push closest to the line where the baby Universe dissolves /disbands from the mother Supreme - Universe.

Due to the diversifying roles of the types of energy during the development of the Universe, increasing amounts of some densest /thickest forms of energy progressively push towards the initial separation line of the baby Universe for forming the 1st-barrier of the 1st-energy-belt (drawing EU72b).

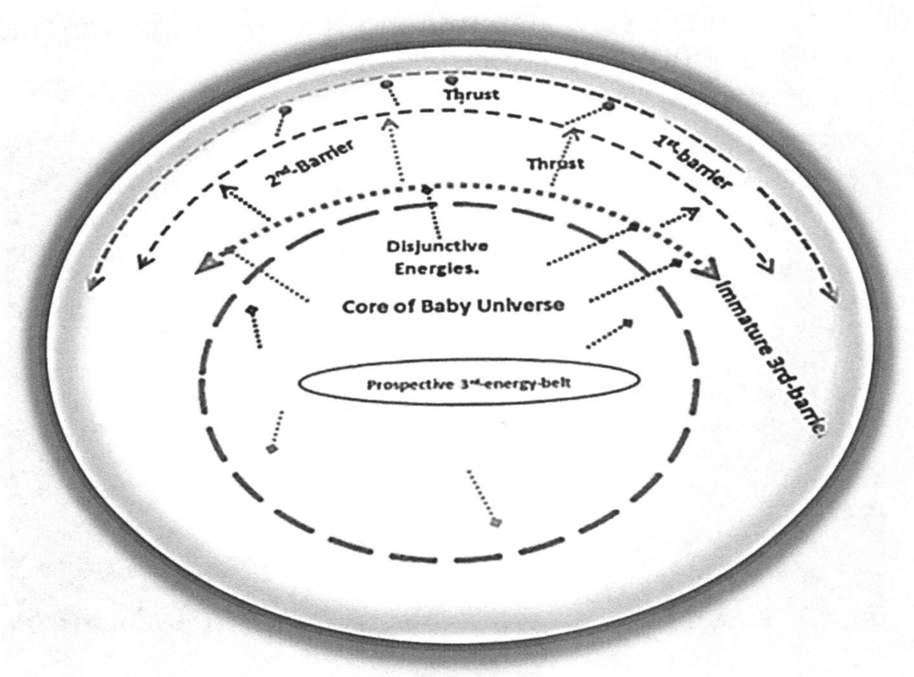

EU72b

→Also, during that exercise for the preparation of the 1st-barrier areas, regularly thrown away, discarded, or excreted many lesser viscous /dense dynamic energy elements, Elemental values, and other forms of masses towards the interior side of the progressing Universe, which ultimately resulted in the individuality of the 1st-barrier of 1st-energy-belt. And so comparably could maintain persistency with the highest viscosity than the rest of the energies in other zones throughout the Universe.

b. In the meantime, some centrifugal dispersals from the prospective 3rd- barrier participated in constructing the 2nd- barrier, which ultimately sets with **vacuum-like** or **null-effect highly flexible fluid energy** (*drawing EU72c*), extremely fringed by the individuality of the 1st- barrier and the 3rd- barrier.

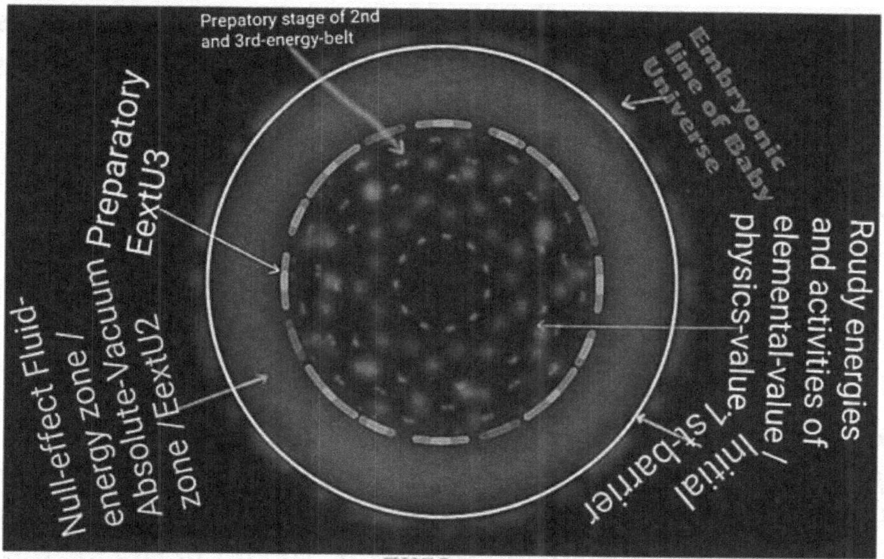

EU72c

→In the growing Universe, due to prevalent turbulence, irregular motions have inflamed more and more with some of the mysterious **time-constrained** *(here **time-constrained** justified on energy levels' momentary energy-Times)* physics values tool like the 'act and effect' that led to the utilization of the word senses like atmospheric physics, temperature, pressure, resistive and so on applications, which tended to possibly, fastest flick off from the congruency of the sovereign status of constricted energies in the prospective 2nd-barrier since it situated closer to the early evolved densest 1st-barrier next to the state-of-Absoluteness, and comparably less dense 3rd-barrier closer to the growing **delicate and multifaceted** 2nd-energy-belt.

*(Here, the word sense **'delicate and multifaceted'** justified following the eternally altering roles of energy-physics, beyond some particular ways of discernment, like the classically involved observable sense, which is primarily valid for applied physics and entirely unlike for direct implication to bio-physics and realizable impact on living organisms.)*

→However, inside the 2nd- barrier, all those physics-values' variations gradually harmonized and ebbed down due to nullifying activities, diluting and absorbing internally developing **vacuum-like** or null-effect highly flexible fluid energy, and resulted in becoming the strong challengers from all possible constitutional disturbance that may arise in the Universe, besides maintenance of the perfect balance in coordination with the state-of-Absoluteness of the Supreme - Universe through the 1st-barrier in coordination with the 3rd-barrier of the 1st-energy-belt of the Universe.

*(By the way, the word **'vacuum-like'** is employed following the many familiar perceptible conceptions that the observable items like matter, materials, and likewise miraculously come from and can freely perform their diversifying roles in the three-dimensionally satisfied word-sense like vacuum or nothingness status.)*

c. Moreover, due to various physics-values effects, all those suppletive from progressing 2nd-barrier and new arrivals from the interior of the immature Universe begin to line up at the periphery of the 2nd- barrier for initiation and configuration of the 3rd- barrier of the 1st-energy belt.

→Supplementary way to say that consequences of maneuvering towards saturation of integrant and shaping /structuring of 2nd-barrier, it also dispersed many residuals /suppletive at its outer edge towards the prospective 3rd-barrier, in addition to the compulsion due to arrival of some new-new classifiable energies including energy-masses from the interior of the baby-Universe that assisted to preparation of the 3rd-barrier is the baby 2nd-energy-belt (*drawing EU72d*).

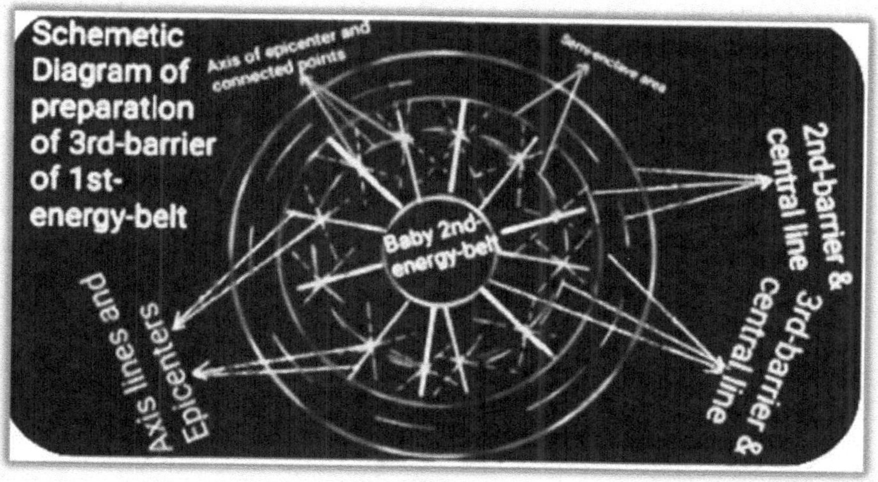

EU72d

d. Besides, all three barriers of the 1st-energy-belt perfectly act as a shielding barrier of the Universe inside the womb of the mother Supreme - Universe.

e. Upon the formation of the 3rd- barrier of the 1st-energy-belt, the remainder of dispersal energy, energy masses, and accumulated elements, or altogether, can say the rest of the dispersed residual thrown toward the interior of the baby Universe for the preparation and progression of the 1st-energy-belt.

f. Whereas, inside the immature Universe, all the energies persistently emerge with reckless and limitless activities that are usually realizable on some particular dimensional assessment.

These reckless and limitless dissonant roles of residual energies are due to the loss or proprietorial changes of necessary potency of operating essences or energy elements in any assumable unit-value mass-energy, resulting in lowering of the effectiveness of elemental values that debilitate to maintain healthy secureness from earlier status like the provisional purity status of unit-value or the primary stage's energy of the baby Universe. And, even someone can think more deeply about the calm and quiet units of absoluteness-state Pure-energy in the Ture Singularity (EU72e) position of the Supreme - Universe.

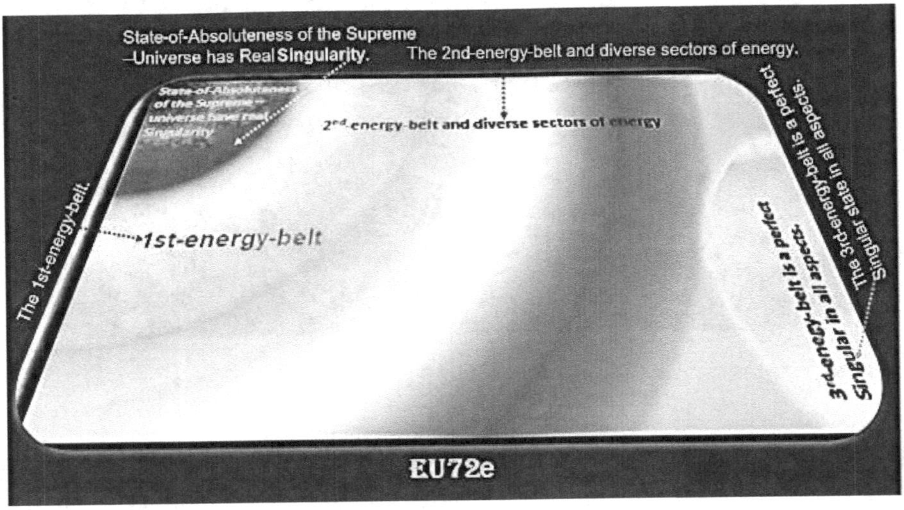

ii. Application of word-sense 'inflation' and deflation /contraction on the point of energy.

In this matured Universe, inflation /expansion of unit value energy may clarify applying the sense 'evolution' of the elementary level of energy elements considering appraisal or assessment of quantification from any unit-value primordial energy and its integrated energy elements upon gradual enhancement of diversity, and consequently 'devaluation' on the point of quality or performance of elemental-values than the counter values' that is comparably upper and lower status units of energy and their integrated energy elements.

→ *(Here, one thing to requote on the aspects of the entire amount of energies of the 2nd-energy-belt nothing can justify as True inflation and deflation, which frequently apply through the delimiting perceptible sense of bio-intelligent beings.)*

a. Some technical view.

For clarification, at this moment, the unit-value primordial energy assuming represents a constant with its componential, that is, energy elements of upper and lower positioning diversity to some dimensionally limiting measurable strategy.

→Besides, the devaluation of the quality of energy or integrally sets energy elements may technically be defined following the formulaic 'mc^2, and say, when some amount of energy justifies the independent feature, then it will represent an envelope, meaning that the envelope will shroud the area like: -

- ▶ **'^2 or ^<2>'** with some form of mass signifies with 'm' and the motion deriving component (c).

So, whenever counting on the sense of devaluation from a reasonable standard point, means the compositionally integrated energy elements of the mass (m) and the exclusive position of (c) will gradually devalue from the standardness due to consistent loss from the standardness of mass-energy (mc^2 or mv^<2>), means gradual dismissing /releasing of elemental values in the form of energy-signals, consequences with degradation of working essences (*that is, types of energy elements*) of the unit-value mass-energy, retaining only the 'c' some level of motion-element at a final thinner state of the mc^2 unit, before dissolving in ostensive Purest-energy, utilize or absorbed by some other form of mass-energy.

b. The ostensive Purest-energy seems ineffective and irrational in all aspects of 3D way measurement systems through the biologically established intelligent sense since the ostensive 'Purest- energy' is the thinnest form of energy and only the inaccessible eternal base or constant for all other forms of energy and objects in the matured 2nd-energy-belt, besides its (*ostensive Purest-energy*) neutral and isolated performance with the extremity of dynamical roles.

c. It looks like observably discerning inflation and deflation is dominant to the human conception besides the preponderance sense of mystery acting as a superior position like a constant for most perceptual measurements.

→But, on the point of energy, the word-sense inflation and deflation may justify as averaging outcome while **multi-directionally** (*here **multi-directionally** is not limited to some specific ways of determining lower and upper levels of deviation)* deriving variations of energy quality derived due to persistent activities of integrated energy elements, from a state of appearance of an object or some form of mass-energy delimited to some determinable ranges.

d. (*At this moment, it may remember a little for conceptional clearance following the roles of energy transformation or transmogrification, or applying the same with a known word like 'desublimation' of energy to some state of matter either of solid, liquid, gaseous, plasma, or so on those are yet to invent.*

→*So, for all the circumstances of conversion to different forms /configurations, from 'energy to object' and 'object to energy. For all the cases of mutation, the quality of 'energies' always gets affected, whatever the standing status of objects justified on some parametric determination.*

→*Whatever the reason for changing the quality of an object, the overall quantity of beholding energies by the 2nd-energy-belt remains unchanged.)*

e. The word senses, like flexibility, pliability, elasticity, etc., in general, are used to understand the properties of objects.

But from the viewpoint of energies in the 2nd-energy-belt, I think those perceptible words are convenient to convince and understand some levels of persistent diversifying roles of energies in the 2nd-energy-belt.

→Accordingly, beyond the observable sense, it is reasonable to justify that energies can be elasticized or expandable, recurrent in appearance, and have the most versatile properties compared to the containing 1st-energy-belt & the 3rd-energy-belt.

f. Indeed, for all the above cases, ostensive Purest-energy consistently maintained an unalterable value or status in the matured 2nd-energy-belt.

→Likewise, concisely say that the word senses expansion and expandable, contraction and compressible, etc., those advantageously applied for energy, is unlike the accessible objects and objectual products, which are determined regularly through the living systems of bio-intelligent beings, remain controvertible and bear with many enigmatic points.

(*Here, one thing to remember is that energies are everything, whatsoever dimensionally determining observable sense satisfies someone as inexplicable beginning and ending of everything.*)

g. Further, from the viewpoint of energy, the Universe, its holding energies, and the autonomous 1st, 2nd, and 3rd energy belts, the word-sense expansion or expandable, contraction or compressible, etc., nothing more than some state of satisfaction to the consciousness of living beings.

So, those types of words are inappropriate for direct application for energy. Instead, the value of units of energy disperses or gathers by improving or weakening on the point of quality and property of observable and non-observable items.

It also means that the energy levels' elemental evolution and devaluation initiate in-depth engrossed energy elements in any unit-value mass energy frame since at energy level loss and gain consistently going on to secure zero-status-stableness as a state of appearance either as mass-energy frame/s or their combining desublimated object in the 2nd-energy-belt.

h. From the aspects of vast energy-Ocean, no query is justifiable concerning the observably applying word-sense vacuum, nothingness, vacant, empty, etc. Rather, some form of effective or ineffective forms of energy (*either of mass form or free-flowing*) occupied in the place of such notionally impenetrable zone when explaining the functions of observable and non-observable objects in the vast space of the 2nd-energy-belt.

i. Upon maturity, overall energies of the 1st-energy-belt occupy appropriate energy elements and functioning elemental values to retain balancing activities and sustenance of provisional neutrality or absoluteness status like the Pure-energy.

Besides applying the word sense like a geographical advantage, the 1st-energy-belt got more leverage than the 2nd-energy-belt and the 3rd-energy-belt since it (*1st-energy-belt*) is closure to the Supreme - Universe, which is a non-physics values' state-of-Absoluteness position. Also, the 1st-energy-belt determined its security through the combined efforts of three shielding barriers (EU72f & g).

j. On the other side, any assumable unit value energy in the 3rd-energy-belt reached its extreme dilution, that is, the justifiable motion element arrived at the ultimate finest point, retaining only one of its kind, that is, the motion of the highest degree and neutrality (*ineffective*). Hence, there is no possibility of applying the sense of expansion, dilution, contraction, etc., those primarily applicable for three-dimensionally discerning observable and non-observable items.

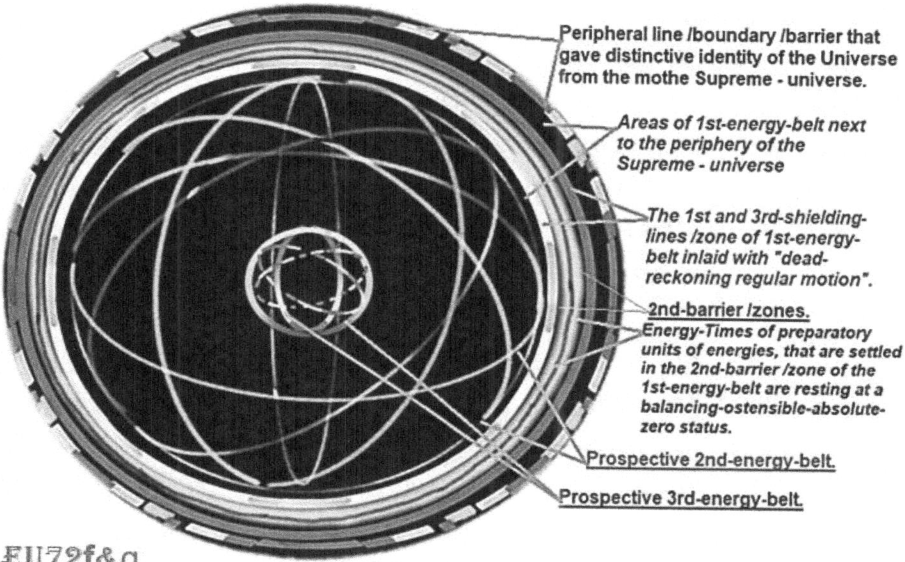

.EU72f&g

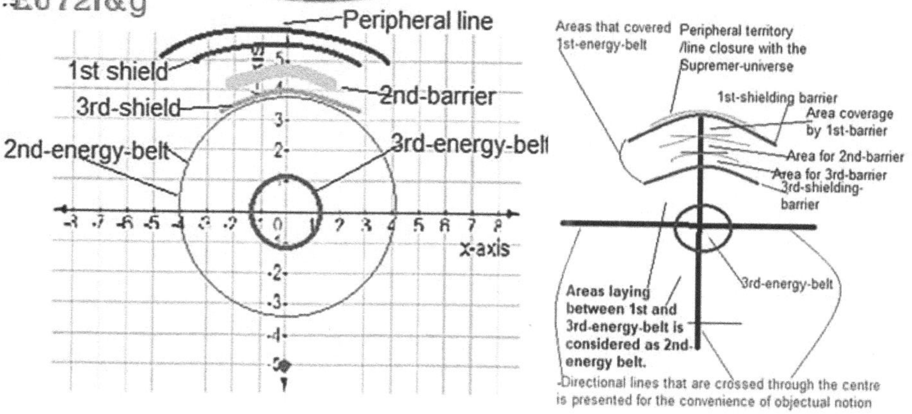

k. Besides, upon maturity of the Universe, the 3rd-energy-belt may be justified in an observable sense as absolute zero positioning since it brimmed with the Purest-energies having no physics-values diversity other than extreme motion and neutrality.

Also, the barriers between the 3rd-energy-belt (*zero-dimensional or 0-D role*) and the 2nd-energy-belt (*multi-*

dimensional or multi-D physics values balancing roles) firmly secured their independent identity.

→ *(By the way, it is to recite that instead of creating too many new words to explain energy levels works, considerable readily available word senses that seem appropriate to comprehend through a variable constancy of bio-intelligent species, conveniently used to write the books for ease to understand the grass-root levels' works of energy and preparations of the Universe.*

So, at some points, it may seem illogical to comprehend how energy levels work. But following present sociological circumstances, those words befit to understand the works of objects, which usually judge through the specific way designed measurable systems. Though, I think It will be reasonable to understand even by the simple biological beings.)

Chapter 20

In Search of The Reality of Life!!

The below points are in responses provided as answers to someone's questions or came to mind but not answered. It is recollections of those responses and clarifications if these explanations satisfy some unsolvable quests while existing as an intelligent living being.

i. There exists a quantum paradox that questions the state of being alive or dead.

Whenever referring to quantum incongruity, if someone justified that 'Energy' is the whole and sole of everything. Then there's no complication and doubtfulness for making faithful roles and interlinking relationships between biological and non-biological nature, the behavior of matter on the particle level and its classical application, and energy.

→I understand, "So far, there is no such technological development in this Globe that can determine the physique of a corporal system alive after declaring its physical appearance as dead, which supports the reality of classical physics for specifically designed bio-intelligent species."

Although, from the classical viewpoint and from the ancestors to the modern scientific era, many human beings this way or that way acknowledge or have faith in death and life entangled by three-dimensional (*3D*) depiction. It is somewhat geometrically definable interconnected triangular lines (EU73) as past, present, and future life.

These delimited triangular ways of spiritual faith deny the judgment about the reality of energy levels physics and determine death and life following only the physical exuberance as living appearance.

That triangular way of existence of life almost takes the superposition of many human minds and reasonably indicates the quantum paradoxical relation of one's existence as present life, future impression, and past life.

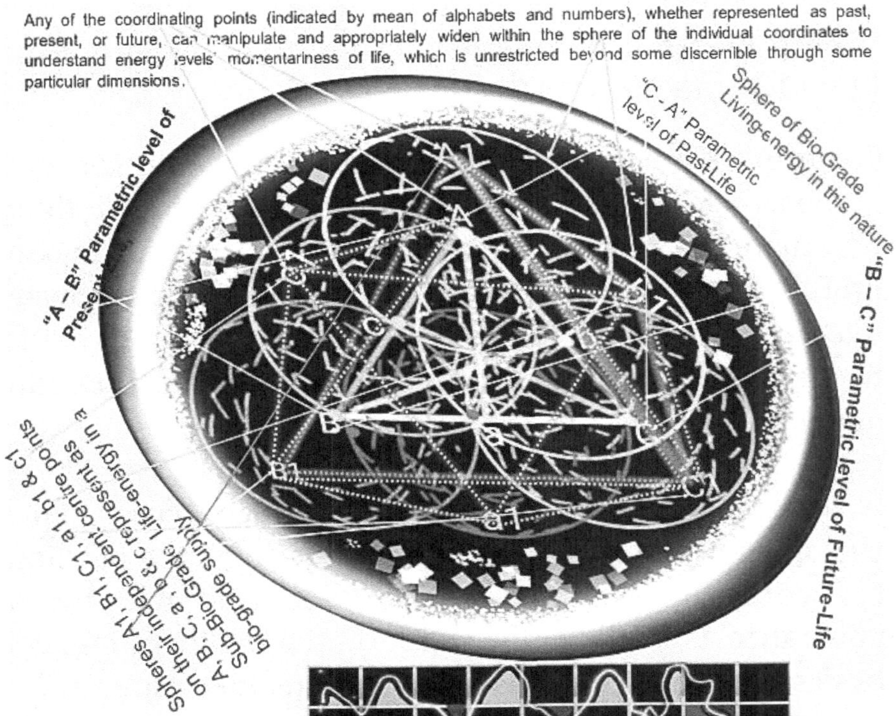

The drawing **EU73** is already defined in Volume 2 of Mythology, Philosophy, and Science about the three stages of life as Present, Past, and Future, which can be triangular ways entangled in circles.

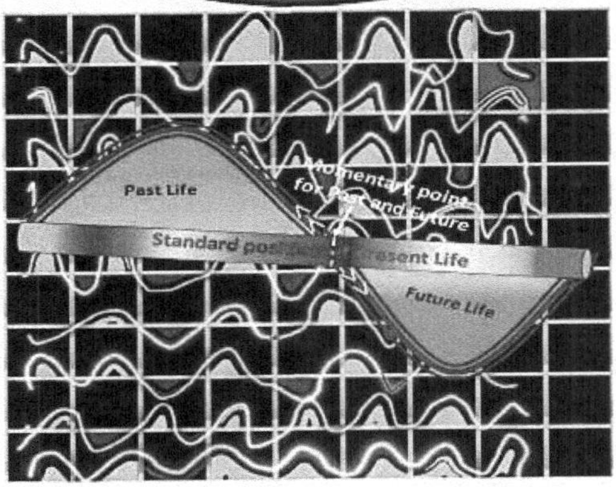

→It may also mean that any be-living physique enjoyed the past life before someone perceived or witnessed its present life (*present life is reasonable constant value as noticeable reality at a place of existence or appearance*) and then will step towards future life upon death.

Somewhat, it indicates the paradox of quantum superposition that the probability of the life of a physique may exist simultaneously both as alive and dead when judgment analyses only on naturally gifted pre-set 3D measuring system of living beings.

At this moment, for clarification, if I go a little with remote sense, which is beyond the classical way presented 'life and death' that follow only the body mass or physical property as a state of appearance.

Then, according to the quantum perspective, it may be reasonable to suggest that the existence of a living body mass extends beyond the perceptible limits of contemporary life. It could mean that there are multiple dimensions to the life of a living body, including an indiscernible unknown state of vitality beyond the limits of life after death.

Further, assume a little more believing an animated module designed for human purposes based on classical physics composed of some matter, chemical particles, or life-deriving components *(i.e., living molecules or mass-energy frames)* like Artificial intelligence (AI). Accordingly, any implementation concerning the rules and regulations, besides authoritative roles, will be justified as non-natural.

Indeed, general conception will grant or may represent with the excuse that 'AI' is an animated machine and human creation, compositionally integrated with microparticles, which is undoubtedly originated from energy and capable of handling directly or in a remote manner, and is unlike (*here, the word unlike is justified since analyzing process performed based on 3D pre-set value*) the proprietorially pre-set acceptance as a natural creation like living beings and compositionally occupied living form of energy that recognizes as chemical energy in biological science.

In this regard, it may be well acceptable to many about the limitation of bio-intelligent beings to understand their findings and works of various states of matter beyond the accessible limits as a state of appearance biologically positioned human species, besides entertaining life by applying the limitation of perceptible sense as a state of appearance like living beings comparing the innumerous holdings in an observable Universe and its proper management.

Accordingly, explaining in a logical sense, physically recognizing the corpse of a human or animal can make it alive by providing the healthy maintenance of constituent energy elements, regularly equipping the requisite /essential amount of life-supporting energy source for maintenance and revitalization of the fatigue or inactive body mass, which once nourished by the mother nature of an inhabitable Globe like our bio-Globe.

→Further, technically, from the perspective of a specific kind of life-supporting energy (*that is, bio-energy*), mother nature always tends to save and secure her creations by providing shares from its earning sources and necessary mutating the structural importance at her best at energy level (*since energy elements are the most flexible and easy mutating invisible items comparing to easily accessible items through specific way designed bio-beings and their supplementing aids for recognition*), comparing the matured mass-energy frameworks represent as independently conscious living molecules or their union as physique who always tend to violate following unnecessary and exaggerating practices for this or that reasons inviting the fatigue or even death of an aliveness body mass as early misfortune.

a. As proof of a non-physical entity, drawing EU73 also represents a general inquiry about the future existence after one's physical appearance has diminished from the present life.

For example, if someone comparably assumes that his physically appearing life expectancy is a point-to-point, mean three-dimensional movement from 0 beginning or birth to 100 years of the attending instances of diversified events, in support of integrated molecules.

→I think an agreeable acceptance to everyone is that 'life' keeps moving and altering momentarily on time, from any point of present status that may be invisible and unacceptable to many when judgments resolve through the constancy of preset value of estimation delivered by the conscious brain processor of human beings like 3D summarized regularly changing Past - Present - Future of total 100 years events while existing.

{(*The 3D summarized regularly changing Past - Present - Future may depict as the methods of retrieval cues, someone who can remember past events stored in memory, while the Future is unknown. It means casting aside the past following the momentariness of the present to step into the new* (somewhat reserved in memory)*, which once acted as a Future.*

Here, 3D ways summarized regularly changing Past - Present - Future of a total 100 years of events while existing may be little elaborated following the sense of energy levels'

Time-dilation of an average body-mass or mass-energy to be utilized from a presumptive differential constant or stationary value as 'Present' while warping the Past periods of existence for the projected Future Time.

So, following the mentally discerning occasional events, the Past measurement may appear slower and relatively stepping fast (or vice versa) towards the Future based on the constant Present occurrence.

Remember, energy levels 'energy-Time' or more in-depth elemental-Times of energy element which does not care about any event and particular ways of discerning temporal Times.

Instead, energy -Time is unimaginably instantaneous with the variation of any standardness of energy-value or in-depth elemental-values' of energy element of a unit-value mass-energy or a mass-energy frame.)}

Also, someone may accept that the integrated molecules are active upon acquirement of the appropriate amount of energy to determine their presence as a zero-status-stableness in different organs of a physically appearing living system, so the overall mass-energy frame of a living being displays healthiness.

At this point, it assumes that the life of an integrated molecule is much shorter compared to their unitedly representing living being.

Accordingly, from the perspective of energy, it is convincing to say momentariness initiates in-depth energy levels of the molecule when justifying that a molecule is a life-orienting mass-energy frame, occupying some amount of energy and integrally active due to the regulative performance of the constituent energy elements in a reasonable surrounding circumstance.

→Based on the above analysis, instead of a partial judgment of the overall body mass of a living being who witnessed billions of molecular events as life for 100 years, that total expectancy of life can shared with all individuals' molecular momentariness or more in-depth energy levels' momentariness of molecular mass-energy frame and accountable their 'life' three-dimensional moves as past-present-future within the periphery of facilitative energy resources, which depicted in the drawing (*EU73*) where any of alphabet either capital or small in group of A, B, C may consider as present and it's relevant coordinates are the past and future acting within a spherical periphery that indicates facilitative surrounding.

As indicated in drawing EU73, any of the coordinating points (*denoted by mean of alphabets and numbers*), whether represented as past, present, or future, can manipulate and appropriately widen /augment within the sphere of the individual coordinates to understand energy levels' **momentariness of life**, which is unrestricted beyond some discernible through some particular dimensions.

Anyway, concerning the sense of '**momentariness of life**,' It's worth noting that the fleeting nature of life's emotional experiences pales when compared to the energy levels that drive the natural and physical world. At the very core of creation and even in the most fundamental particles of matter, energy plays a crucial role in determining the momentary properties of any given system. It is especially true when considering the relationship between mass, energy, and its integrated energy elements, the actual functioning essences of a mass-energy frame.

ii. Talk about the physically endowed words like dead, decay, and so on, which are inappropriate when considering the aspects of energies.

When someone talks about Energy, the true sense of expression through the word's death and decay of biological physique does not arise. Instead, it is simply a metaphysical alteration of the once corporal system to Energy or a different physical entity.

Energy is never exhausted but transforms one state to another upon the end of status as an appearance either of a living or non-living mass-energy frame *{(mc^2) or (mc^<2>)}*.

Unit-value bio-intellect is independently shaped by millions of molecules structurally developed from bio-grade energy sources, which, in general, are available in this bio-nature.

So, we need to consume healthy or appropriate foods that indirectly provide the necessary Energy supply to the constituent molecules and their unitedly operating unit-value big physique.

a. The next point is the traditionally trusting and celestially representing the soul of a lively physical system, generally applied as an external purview but essential celestial source in-depth of an individual human body of any mono-state human being, so beholding an eternal heavenly dignity.

However, when justified as the union of multi-types molecules, it is appropriate to consider it (*the soul*) as the natural gift of conducive surroundings that give them (*constituent molecules*) life to stay lively /active when they act as fully-fledged operating components for the union representing as a living physique like bio-being.

→Further, when molecules come for unification to give accumulated shaped as bio-intelligent beings, they become entangled and dependent on the whole unions or all molecular cooperating efforts to save the unit as a healthy human being and even for their individuality as existence.

And, as long the constituent molecular combining efforts appropriately support securing wellness of the soul of a unit-value human being, it got authorization to enjoy the essence of life under a flagship of living status.

→So, the soul of an individual human being, and likewise other unit/s of living species, is due to the contribution of integrated millions of molecular supports.

Accordingly, the soul of an individual bio-physique occupies the specific property, on the point of heavenly purity that may apply, somewhat similarly to the faith that usually sermonizes through different ways of expression, like the "Paramatma, Almighty God, and so many distinctive ways of praying for divine the Supreme Soul."

b. Furthermore, phrases such as Karma, Fate, and Destiny are typically associated with religious beliefs. These terms attempt to justify the summation of an individual's actions in the present, based on the assumption of their previous states of existence that supposedly determine their future due to an unknown course of activity.

However, as per general acceptance, the appearance and existence of an individual are primarily a result of the consequences of their past activities, which subsequently drive their physical system in either a positive (good) or negative (wrong) direction and can be regulated through the developed intelligent brain processor (*in essence a 3D measuring practices*). Further leaning with the above, the same restrainment can apply for narrow and wide-ranging comparisons, from the smallest living organisms without physical brains to larger ones' distinguishing brain-like controllers.

Therefore, following the socio-culturally developed spiritual faith, besides applying the technical sense of exaggerating gravity, that is, strong attraction for materialistic life loomed due to native desires, greed, etc., restricts the voyaging of the soul on its free will when a physique is alive or upon exhaustion from a living carnal system.

→One thing to remember is that desire, greed, and so on, as carnal attractions, may seemingly be permanent due to circumstantial influences following the native elegancy that may satisfy some points of physical requirements.

But from the viewpoint of Energy, everything is provisional or momentary and has some term-bound and time-bound balance.

Accordingly, expression through the words Karma, Fate, Destiny, and so on religious belief is appropriate while enjoying liveliness in a place of existence in this bio-nature and distant nature.

[{(*However, many people ignore the existence of supernatural activities that occur around them, whether predisposing sensing faculties* (like the eye, ear, skin, nose taste) *that are specialized to analyze temporal introductions can detect or remain obscured entirely to the period of the overt state of physical appearance as personal pronouns like I, We, You, Me, He, She......, etc.*

So, knowing actions following worldly influences for the temporal requirements does not verify the micro-level misery that acts in the depth of molecular levels, whose individual positions for different organs functionally simulate and enact on the brain processor that may disturb administrative roles for the entire physique.

Therefore, materialistically deriving virtual truth may disdain long-term prospects for individual physical dignity whatsoever ideology dominated as an overt or covert state of physique or both, sustaining supernatural faith.

Also, the involving tiny molecules that soft-soaped due to excessive worldly items while participating for short term (compared to whole body mass) *in an organ, either those are influenced, seasoned, or constructed from materialistic energy signals* (one of the 3D discernments) *to perform fleetingly negative or devil roles while invited due to energy levels persistent physics values activities can never avoid repeated appearance as a long-term misery.*

{In this regard, a little more may clarify following the philosophy of Hinduism that the soul of living beings always tends to maintain holiness in the hope of meeting or absolute integration with Paramatma or Supreme Soul,

whatsoever the body mass straightforwardly attracted by the gravity of lucrative worldly items. Accordingly, the soul of a momentarily appearing tiny body mass or mass-energy frame (mc^<2>) bears the same tendency.}]

From the above, it is easy to understand that even though the Energies are everywhere. However, the soul or deriving essence for the functionality of living Energy has distinctiveness, even though it needs exclusiveness on the point of qualities to travel remotely on **free will** due to the prohibition of superiority or inferiority following the principle of Karma, Fate, Destiny, etc., or the working activities as a state of living physical appearance.

{(*One thing to quote is that the physically dignified superiority and inferiority may not perfectly work on the point of energy when religiously believing Karma, Fate, Destiny, etc., that follow the energy level's divine accreditation since visually justified* (that is, 3D ways discernment) *physical works and mysteriously energy levels effect, which may suggest either way.*

(*It means when a man is recognized as a hero following only the co-occurring circumstantial materialistically influencing role, but when the circumstances superseded, the earlier declared heroism may turn to cravenness.*)}

iii. 'Nirvana'

The celestial word senses like Karma, Fate, Destiny, and so on, conveys the practicality of a living body mass whenever endured a position in a recognizable state like a human being and enjoying the materialistically diversified inhabitable planet, besides the 'Nirvana' depicts where the physicality of livings has to come to a transcendent state, believing it as a paradise or total serenity from the diversified role of animated physical impression.

The supernatural word sense 'Nirvana, in essence, is regarded as overcoming all carnal desires by subduing the materialistically occurring passion that persistently disrupts the True or long-term satisfaction when existence as an active life with surroundings influential temporal avaricious items. Besides, the term enlightenment became familiar in the Western world around the 19th century as the concept of supernatural Reality.

So, divinely representing Nirvana is possible upon the total loss of consciousness and release of an individual soul, closing down the chapter of all differentially deriving suffering, which happened due to disappointment to ephemeral amusements while existing in a physical appearance.

In brief, I may go with the profundity of energy levels' philosophy and try to illustrate in a brief and reasonable sense for common understanding about this celestial teaching 'Nirvana.

(*One thing to remember is that the reality of energy is eternally holy and has no elegancy, dirtiness, or any directionally judicable partial discretions. Those senses are mainly applicable to the physical status of conscious living beings, primarily concerned with physical well-being, often believing that the purpose of life is to indulge in material pleasures only. So, somehow, some opportunists try to attain satisfaction through conscious and unconscious emotional actions and sometimes by manipulating reality with words to solve carnal purposes.*

At this point, it is significant to recall that energy is a fundamental source of everything and eternally sacred in all aspects, whenever and wherever it emerges as a physical state. And can never be influenced by human emotions, biases, or other forms of living physique or mass-energy frame.

However, energy levels' eternity of sacredness changes due to diverseness application and derivement of immense causal factors, and so enactment to the physical judgment, which discern through some physically deriving consciousness that can distinguish as right, wrong, or any other ways of discretions limited to some materialistically presenting dimensional avocations.)

It is acceptable that the mythological word Shakti and the Scientifically introduced Energies are the same since both are the source for all applications and even the essential parts of whatever the bio-intelligent species are directly or indirectly capable of perceiving through the structural delimitations.

From the viewpoint of Shakti (*in Hinduism*) / God (*socio-Religious faith*) /Energies (*scientific acceptance*), the deep-down source of all differentially evolved physical systems, diversity, and even a minimum state of desirement became necessary to some limits for survival due to daily requirements. Accordingly, irrespective of human beings, molecular systems are also adept at struggling to preserve their status as visible or invisible appearances.

But upon the death of a physical state, or cessation or extrication of the soul, the importance of Nirvana emerges, and all the past physical activities of the exhaust physique represent a transitional (*time-bound*) condolence to the consciousness of others.

Accordingly, it is logical to say that spiritually presented Nirvana /Enlightenment is significant for the teachings of human beings to improve one's self-control and consciousness in balancing the order of the surrounding atmosphere while governing the union of billions of outgoing and new arrival molecules and their diverse requisitions in a lively physique.

In this regard, a little more may clarify the above and say that intelligent species acknowledge that a living physique formed from organic compounds like carbon or non-metallic chemical elements, which is the essential part of life besides a few other fundamental particles like hydrogen, oxygen, nitrogen, and so on. So, it may justify that those typical elementary levels of life-orienting components are constitutionally unique for all bio-living species.

In the modern era, I believe most intelligent beings will accept that Energies are the primary source for everything regardless of classical particles, chemical particles, quantum particles, etc. So, this biochemical directly may consider bio-grade Energy as one of the fundamental sources for living in this life-supporting bio-nature.

Now, it is logical to assume that to some low-level extent compared to a high-level human being, and say, whenever a molecule gets a stable position in support of surroundings, it represents an independent feature. Or the same may justify that the individual personality with consciousness about reserving their (*molecular*) appearance as a module of the living unit they design to perform, or simply that the module as living molecule term as a zero-status-stable mass-energy frame (mc^2).

Hence, a living physique of human beings constitutionally represents a union of actively performing multiple molecules, and a living being or physical state of consciousness is the outcome of integrated molecular offerings. The central processing system, or the brainbox, mainly acts as the judiciary system for ratification and the disciplinary regulator, whatever derivement arrives directly or indirectly due to internal and external physics values application.

In general, true consciousness of a living physique develops in a surrounding atmosphere due to the momentary interactions experienced by the deep-down molecular system that finally processes and is accepted by the central brain processor as a human being, in addition to some genomic contributions from the predecessors' reserved rights (*that reserved rights carry on generation to generations unless and until mutated for some reasons*).

A little more say, as an independent human being or the union of molecules if someone imagines that as a whole, it represents a Zero-status-stable positioning sack of bio-grade Energy under the parametrically definable conducive surroundings. Then it can justify that depicting sack-of-energy as a human being, constitutionally diverse with multifarious sub-class of bio-grade Energy and operating components (*energy elements*), occupies differential properties. It is due to the energy zone that supports visible and abstruse life, constitutionally occupying diversity in all respects.

So, it is reasonable to assume that whenever a molecule gets a stable position upon obtaining the support of the surrounding atmosphere, it represents an independent feature or independent personality with consciousness regarding quality to preserve its appearance as a module of the living unit they enterprise to perform.

Moreover, molecules congenitally organize based on their occupied qualitative energy components gradually set up deep down as the energy elements in unit-value bio-class energy before shaping and getting dignity as Zero-status-stableness statues.

The molecular identity, rank, degree, or pride may currently justify based on the long practices' mathematical arrangement and mightiness of number integers they (*energy elements*) set in orbital manners, keeping the motion-deriving motion element at the core position or 0-center.

Hence, an overall unit-value life-orienting bio-living mass-energy frame that befits to present as a human being intrinsically and individually represents a union of multiple molecules, actively performing so the proudness of an individual human being, including the consciousness is the collective contributions of integrated molecules.

By the way, It may recollect that the 'soul' of a living being prevalently acts as a memoir of events when there is aliveness of the physique or body mass since the 'soul' natural ways got regulative authority besides its intrinsical characteristics as Truly with neutral functionality.

Hence, justifying beyond the temporal aspects of living and plunging deep into energy levels, It is reasonable to accept that the teaching of 'Nirvana' has a supernatural connection to all physical states of living in any inhabitable nature and the guidance about the best management and control of the suffering of individual intelligent beings from materialistically deriving exaggerative attraction that may lessen or may allow overcoming many miseries of a living physique in a place of appearance like this alluring bio-Globe.

So, following the practice of 'Nirvana' or a better approach for metaphysical management of a state of living physique, the possibility of overcoming the three-dimensionally deriving physical avariciousness.

iv. Herewith are Some points about Free Will and Free Will Action.

The word sense, free will indicates a probable authority and ability to make infinite choices and take actions as a state of independent physical appearance compared to the vastness of the Universe.

Indeed, if someone thinks of individuality as a physical state like living beings and many others, always reserve some personal rights when a living status in any nature and the naturally affiliated society, authorized to function and manipulate to some degree on free will since everything has to abide the rules provided by those caring natures of occupancy and traveling surroundings.

(The above point is a somewhat modified state following the essence of diversity, originally a follow-on gift derived from the great grandmother Supreme - Universe, so instead of reserving the innate Pure-status like units of Pure-energy, any physical status minimally could maintain the unity of constituent energy elements whenever and wherever some amount of non-Pure energy represents as mass-energy zero-status-stableness status in any atmospheric condition of the energy-Ocean in the 2nd-energy-belt.)

→So, it is comprehensible that the **'free will action'** on one's desires is impossible. Instead, I think it may be appropriate to apply the sense of **'free allowance action'** when someone appears as a status of existence bound by the rules of the surrounding atmospheric physics for reservation of individuality.

The expression through the word sense free will is somewhat similar to the total liberty or freedom to make an independent choice.

Indeed, as the socio-culturally developed intelligent species, living beings in this mother nature somehow can manage to perform and act on free will within the limits of sociologically *(that include the socio-economy, socio-science, socio-religious, socio-cultural, etc.)* provided regulations, besides natural rules administered by the settle in nature.

Presenting with a simple example like a person, someone who can laugh, cry, enjoy, or whatever he covets, everything endlessly attempts to do of his own free will, believing he is pansophic. However, instantaneously upon crossing free allowance limits, some unexpected or unfortunate thing may affect the knowledge that seemingly the physically appearing free-will areas.

a. Now look at the molecular levels of a physique where every molecule is somehow and someway dependent on others, almost all purposes following the rules of survival and existence in a place of comfort living, like a human being.

→Further for an explanation, at this point, one can imagine that the bodily system of a living being somewhat represents an observable Universe, and the integrated organs and molecules are the distinctive galaxies and like planets. Then, it is possible to think of all molecules in a figurative living physique dependently entangled with each other for survival necessity and to preserve the physical Universe (*living being*) at a state-of-Stableness in coordination with the surrounding conditional atmospheres (*that is, differential atmospheres emanated due to multiple* functionalities of *the energy -Ocean*).

Therefore, following the survival necessity, from the molecular level to the entire lively physical structure, dependent on and have to coordinate with the diversified consumable products derived from matter and materialistic items in an inhabitable nature like this bio-globe from birth to death.

Accordingly, the free will of one's desires also are restrained following the integrated molecular requirements and also the laws (*rules and regulations*) provided by the caring inhabitable Globe.

b. Now, think about going a little deeper concerning classical and particle physics.

All matter, material, and quantum particles occupy some form of 'mass-energy' whenever they acquire structural value like a frame in some form of shape, size, and volume (*that is, dimensional aspects*), and accordingly, all of those need to follow some parametrically regulating rules of the surroundings for evading external physics values applications like energy levels' pressure, temperature, resistance, etc., otherwise will be disturbed or even may transform instantaneously from any occupancy as the physical state with loss or may exhaust totally to something else or free-flowing energy.

c. In essence, from the perspective of energy, everything within this 2nd-energy-belt is causally interconnected.

From the above, it may be clear and perceivable that everything, irrespective of human being and particle, has some free allowance or parametric limits or frame of rules to act on independent free will and likewise.

Also, think of quantum waves, some form of no-mass energy in the waveform consistently playing throughout the 2nd-energy-belt, article, particle, matter, material, etc., which are all entangled and rhythmically playing.

So, the freedom of human desires to make choices and success is not priorly determined and causes for divine intervention only but has universal interconnection with some energy levels' causal effects.

v. Some points about the word sense mind and its related undertakings to bio-intelligents' living system.

In general practice, the word mind can be justified by the ability to determine the perceptual /mental occurrence expressed through words like mindfulness, sentiment, thinking values, desires, pain, and many more internally generated emotional expressions. Those are the factually componential (*here componential mean, constitutionally participated biological living elements like molecules*) outcomes and requirements (*either whole or partly*) for the physique or body mass.

→Everyone will agree that our physical system comprises trillions of active molecules. And every unit-value molecule may think as an independent operator for reserving the liveliness of our living physique.

During the activities of molecules, they generate some types of signals for communication with each other and subsistence with activities. Those signals may directly assume the energy signals (*instead of emanated signals due to chemical and electrical functionalities*) within the biologically offerable specific bio-energy spectrum.

So, from a molecule to their union as a group that maintains individual sections like heart, kidney, muscles, etc., of a physique. All those organs somehow managed to deliver overall generated signals directly through neuro systems to the brain box due to applications of human body physics.

→The mind and its different undertakings are the best internal communication and natural carrier for all individual molecules when entangled with a union of physical systems.

However, the brain acts as an analyzing tool like a computer processor where incoming signals are analyzed and evaluated for feasibility to think nearer or distant products and deliver in the form of 'thinking values' data signals of factual or imaginative.

So, the mind, which in some case function like conveyance, proves its perfectness as well as inefficiency due to the source of circumscribed origin, compared to many other energy products in the Universe.

→Here are some concise points concerning the brain and mind of a human being. Technically, for physical functionality, the brain and mind are correlated even though they seem to perform their independent function, whatever traditional discretion accepts as mono or multipurpose requirements.

Somewhat like, the brain acts as hardware, and the mind is software that collects internal information from the molecular stages of a physique and delivers it directly through neural systems, also indirectly using energy levels physics to the complexly designed central processing brain box.

This brain box or central processing system then judges for action, either conscious, subconscious, or unconscious mind or reserves as a possible retrievable condition for future use.

Indeed, healthy mind signals can travel faster in one's living-physical world. But it does not mean that the mind signals as a datagram (*like the telepathy technique*) capable of challenging all the times of external atmospheric physics and overcoming energy levels' fastest performance.

vi. It is meaningful to understand the concept of civilization and its technical importance for humanity.

Following the acknowledging interpretation concerning 'civilization' means the process and sophistication of human social and cultural as well as organizational purposes.

However, contemporary eras' complex societies where the human species augmented their sociocultural relation far wider throughout the world than the earlier conservative societies that restricted due to some purposes for small organization/s.

Accordingly, the collective humanity and benevolence quality needs to improve proportionately to match the requirements of advanced contemporary civilization compared to the earlier civilized people.

So, it may be agreeable to someone that the civilization does not materialize immediately or miraculously upon appearance as an individual or their group culture without any natural influence that followed irreversible ups and downs while inhabiting this life-supporting bio-Globe.

→One of the fundamental principles of observable Universes is the trend of simultaneous processes of construction and destruction, even if believe it or not, while conception (*mainly assessed 3-dimensionally*) works and analyzes following the individually bearing Constance either of mono state judgment or the overall group finding since observable ranges delimited in many aspects including the factual works of energy.

(*From the perspective of energy, for conceptional clearance, it may recite that there are many non-observable Universes that the human species justifies following their three-dimensional measuring projections, and so far, recognizes as dark matter and dark energy. Besides other types of unreachable massive and minuscule forms of premature mass-energy and ostensive Purest-energy occupied the entire coverage of the matured 2nd-energy-belt.*)

→From the ancestors' eras, human civilization steadily advanced (*which means averaging headway moves*) following the revision /emendation of the atmosphere of rooming bio-Globe that compelled the olden sociocultural practices to bring changes since it required to coordinate and comply with the naturally available resource for **survival necessity** resulted in emerges into the marvellous modern civilization.

(*Further, to make sense concerning the point of '**survival necessity**, it may mention the popular term 'Survival of the fittest' that refers to the process and mechanism of natural selection, which brings the evolutionary changes following the **conditionality of environments**.*

*This **conditionality of environments** may modernize and easily place for energy level changes of specific grade, which is a suitable resource for bio-living species of this nature.*

So, all living products have to call for physical and mental changes and simultaneously become essential to struggle for survival necessity that allows to adapt in coordination with the energy level changes.

→One thing to remember is that energy is the root cause of all changes, regardless of assuming original products or correlated entities.)

Accordingly, observing the steady advancement, someday the bio-human races may reach an amazingly advanced ultra-civilized humanity capable of taking **superposition** of others' emotional sorrowing state instead of showing only sympathy and some physical assistance following the contemporary practices.

That **superpositioning** practices of living may not be suitable for the bio-living of this inhabitable planet, but probable for the distant worlds of non-bio-grade life-supporting energy, where it may not be enjoyable to the same fictional problems of philosophy, scientific findings, and even the same bio-intellectually developed sociological diversity.

→Technically, emotional feelings and the related gravitational magnitude of this bio-globe, which considerably influences human beings, will certainly not be the same or hardly imaginable through the biologically developed consciousness in a bio-system when considering the distant inhabitable non-biological Globe.

So, the bio-socioculturally evolved many word senses that please or make someone calm from the excitement and bitterness following human behaviors while unitedly living as bio-species, probably find no similar place in a distant habitable Globe since those alien types of sympathetic feelings exist or operate through an affinity, interdependence, or mutual association following specificity and different ways of materialized competency.

→Distant living species may be more advanced than someone (*bio-human*) can imagine following their pre-set observable sense. And those alien living can probably share their self-holding energy as a physical form for others' requirements, which is unlike the practices of bio-human species, those who are always dependent on physically appearing consumable items indirectly to provide necessary energy sources for integrated functional units like molecules, besides always toil due to some form of confrontation for natural and commercial availabilities.

vii. Here are a few points about the notion of aliens and extraneous influences.

The general conception of an extraneous living is somewhat of a physically discernible strange body.

→But following the perspective of energy, aliens may not require any physical way of appearance, which befit to discern through biologically evolved intelligent species delimiting in many aspects to introduce non-biologically evolved alien intelligent species and realize their presence and undetectable influencing signals.

For example, someone may realize the non-physical existence of aliens' impact through the commonly assessable own physical world. If one really can meditate for long hours *(indeed, it may be difficult for the general people to go in depth of meditation*) following heavenly, Godly holy items, or alien of good God or Goddess roles, then realizable through overcoming many predominant **mental sensations** about the non-physical alien with godly influence.

(Generally, ***mental sensation*** *has profundity on physically appearing conscious awareness that evolved and became preset values in the biological beings' systems while adapting in a native nature like this bio-Globe, so it becomes unrealistic when lively bio-body mass or mass-energy frame intercepts some intelligent alien signals.)*

On the other hand, if one can meditate or contemplate for a long or short time like hellish barbarism or some state of the harshest scenario, then the non-physical signals of evil aliens' influence may intervene and promptly impact a brain processor that realizable through the living physical system.

Unknowingly, it may even conquer one's internal angelic attitude and derive towards unusual unangelic rudeness roles that consequentially derive away from a pleasing appearance to the worse miserable course of an individual, and that directly or indirectly may influence the whole society when applied extravagantly.

(One thing to quote is that the perception about the physically or materialistically emerging heroism is not truly the energy level's indomitable potency. Instead, physical strength restrains due to the loyalty of appropriate energy supply.)

viii. The creation and utilization of multiple words have meanings for benevolent, malevolent, and self-centered intentions.

In this regard, foremost for convenience, it may discuss little concerning the purposes of the modernized ways of utilizing words and generalization of how to express and understand between the members of a species such as advanced living beings.

→I think everyone will agree that immediately upon the birth of a human child, the newborn does not know the meaning of any words or specific technique to express by using words for physical requirements, so the caring mother generally realizes through observing the bodily expression, besides non-observable remote signals derived internally by the aborning physique that capable of inducing mothers loving heart and allow some sorts of realization of newborn requirements.

[{Besides, the gravitational perspective of the mother and newborn's direct soft-loving soul or True heartily-loving gravity is much stronger than indirect and purposely generated gravity by the body mass (mc^<2>) of the mother, newborn, and other parental caring distant one.

(One thing to remember, at the energy level, 'gravity' is nothing special, but it is a constitutional claim of a mass-energy frame derived due to diverse roles of functioning energy elements.)

{(*Technically, this direct soft-loving attraction may be somewhat exemplified by the vital body parts of an individual being since True* (not feigns) *emotional roles work deep inside the molecular level and are discernible when they unify as a union, making strong gravitate relationship to perform an overall zero-status-stableness. Then, no one unfeignedly will desire to separate or discard any part* (physical state) *from the whole-body mass while existing healthily as a union unless otherwise, external leverage like the impact of energy physics acts as 3D measuring anti-gravitate roles influence negatively for segregation.*

Further, for example, In the case of a mother and her newborn, soft-loving attraction begins farming while the fetus or an offspring (mc^2) is in the stages of prenatal development, so expected aborning keeps communicating with the mother's body mass (or at this juncture, this body mass may assume as the union of molecules or integrated organs)*, applying internal body physics, which is generally remote to realize and yet to discover.*)}

→Sameway, at least someone can contemplate the primogenitors' expression techniques and more in-depth about their integrated molecular expression or communication for their requirements.

Based on the communication technique, I think molecular communication seems more simplified and honest than the 3-dimensionally manipulated communication techniques of modern human beings, by using words, distinguishing them for different purposes like benevolent, malevolent, and self-interested /self-centered intentions {that is, good or positive, bad or negative, and one's self- centric status or a natural constant (*here, bypassing any sociological materialistic criticism, this word sense 'self-centric' at this point may assume as a natural resonance of in-depth at energy levels practice tending to reserve zero-status-stableness as a state of mass-energy frame physique under the consistent duress of diversifying atmospheric physics*)}].

→Following the above, it is comprehensible why many people advantageously prefer some words that can easily manipulated and applied to make pleasing sentences look forward to solving many immoral purposes, like faithfulness, religious belief in God, Philosophical and Scientific findings for good objectives, etc., which are truthfully appropriate for virtuous /moral interpretation.

But some people manipulatively use those ethical words following the circumstances, opportunely and purposely apply for three positions as good, bad, and self-centric by involving 3D ways computation techniques such as (+) positive or sociological goodness, (-) negative or sociological badness, and differential constant (Δ) as an individual physique to bring some satisfactory state for an individual, and one or more participants and even convincible to many of species.

→Attainment of the truth is a philosophical phenomenon that resulted in the improvement of consciousness from the depth in molecular stages, a constitutional benediction, besides the genomical teleportation, contributed from generations after generations of the human species.

→The living beings in this bio-nature are structurally fortunate and manipulatively more intelligent than many other living species, so human beings conveniently, deliberately, or proudly present themselves differently with heroic word senses for their own and the group's satisfaction.

→Considering the energy and following the componential and structural significance, living beings are inadequate in many aspects, so it is easy to get deprived of many True energy levels' findings due to suppression and wrong maneuvering practices. So purposely and effortlessly, someone can exploit the natural ways preset mathematical establishments in the living system, whereas the brain processor computes all assessable entities unfeignedly.

The independent brain processor attempts to analyze all the possible delivered energy signals, introduced or intercepted by the constitutionally established molecules and the delimiting primary sensory organs, which are the acting testimony of a physique. Thereon, the brain processor interprets and derives the processed outcome and delivers through the physically established components in word and sentence forms that are grammatically sensible and match many synonyms and antonyms linguistic expressions such as True /False, Real /Fake, etc.

Also, those linguistically acceptable words and sentences many moderators of distinctive communities advantageously exaggerate and apply for their purposes that may call for an early disaster of the human species, obscuring the Truth and long-term reality.

ix. The act of survival and its functions impact the physical systems of all living species.

The practice of survival is a fundamental requirement for all living species, as it refers to the ability to maintain life in the face of adversity. It encompasses a range of biological techniques and adaptations that enable organisms to cope with environmental challenges, such as predation, disease, and resource scarcity. The impact of survival on the physical system of living species has profundity, as it shapes the evolution of species over time, influencing their morphology, physiology, and behavioral context. And ultimately qualify to survive for the continuation of life on this habitable bio-planet.

→Our ancestors were once Truly faithful to nature as they depended on some narrow areas for survival necessities, so their method of communication was restricted compared to the wide-ranging modern advanced generation but maintained unproportionate alteration of the self-esteemed essential structural values and 3D measuring methods.

Hence, the deficiency prevailed from the dawn of the human species during the process of evolution that persists even in modern intelligent beings when judgments anticipate some extra dimensions and tend to measure following the restricted ways augmented preset values' configuration in the brain processor, which mainly analyzes through the delimiting sensing signals and administering physical requirements.

Therefore, all analyses somehow become linked with the dwelling nature, which acts as a hidden immutable constant whenever measurement attempts to perform beyond the 3D processing system, maintaining types of deficiencies at some consolidated position for the satisfaction of an individual or the entire species.

→From the viewpoint of energy, survival did not mean a simple purpose of gaining, consuming, or eating and fulfilling the energy requirements of the physical systems.

Still, there is greater competition for accessibility to get appropriate quality and quantity of energy, which is reasonable for any physique in a native /supportive atmosphere.

Therefore, the sense of struggle came into force and slowly became a dominant factor and regular practice in living systems.

→From the viewpoint of energy, philosophy became the by-birth property of the living system, the great gift of mother nature when she processed his exceptional creations from diversifying energy elements and appropriate application of atmospheric energy physics from the bio-grade energy source or components, something considered like carbon-based energy components, which is conducive for the livings those introduces from biological-life operating energy resources.

→Mother Nature is an exceptional scientist who possesses and efficiently utilizes a plethora of energy resources to manage all her holdings, even though she maintains intricate relationships with other energy resources throughout the 2nd-energy-belt since it requires specific types of inhabitable energy supply to get an exclusive form of life-orienting bio-energy for her elegance creations.

→It is important to note that energy has fleeting properties, and the lifespan of biologically living organisms is relatively long while counting many other energies level factors on this habitable bio-planet.

However, when it comes to an overall life-orienting energy in the 2nd-energy-belt, **life never truly ends**, even if the physical appearances exhaust from a state of appearances.

x. Here are some significant points about the ideas of the word sense 'life' and its extensive application well beyond the human species.

By the way, in a summarizing manner based on the general conception and differently defining 'life, which is mainly for biological living following some perceivable sense, so can distinguish the vitality and functional activities like eating, metabolizing, breathing, moving, reproduction, etc., that solve some survival necessities and to maintained some animated roles compared to a dead physique.

{(*Here, the word dead represents contrary to the sense of life since nothing more persistent of socio-culturally accepted temporal living activities of the body mass (or animated roles of a unit-value mass-energy frame) as mentioned above.*)}

→Following the general conception concerning notions like life, death, and the temporal time-frame existence, it retains disputable situations whenever someone intends to raise queries about before the commencement of life and memoir of present times while existing and after death.

Indeed, many people use expressions like "nothingness" or "everything is a miracle" to highlight paradoxical feelings or to justify their complacency, regardless of how they view life, death, or existence.

Over and above the miracle and nothingness, some human beings are atheistic when many other deities have some faith in God, following the natural ways of development and perfection of design to survive in this biologically flourishing exclusive form of bio-energy.

→From the structural viewpoint, the human species' natural ways are delimited in many aspects.

So, the sense of miracle and nothingness, which engenders in a living being's system while the inbuilt consciousness undergoes many unanalyzable events due to the dwindling of componential (*body components*) physics even from the standardness of preset sensing organs that progresses following some natural reasoning.

→At this moment, in brief, it attempts to match the biological living system with Energy.

Possibly, the sense of Energy once was obliviously accepted by human predecessors, and with the gradual advancement of generations, the essence of Energy enlightened and allowed to overcome many darkness and unreality that turn into actuality to modern human beings, besides augmented understanding of its (*Energy*) applications in all aspects of universal Energy level's activities (*that is, imaginable most micro to the large state mass-energy frame following the unique presentation E = mc^2*).

→Why Energy-God!

As and since, Energy has solitary personalities (*regardless of judging as good, bad, and impartial)* in all aspects of universal activities, including the life-orienting mechanisms.

So, Energy can be dignified with the highly holy honor of God or characterized as Energy-God, who is factual for all miracles and consistently undertakes for caring of all the imaginable and unimaginable most micro bit to a massive unit values functionalities, whether someone accepts existence as a physically appearing living form or miraculously disappears after death and never could find out during the period of someone's (*near, dear and distant one)* liveliness or their group declaration as eternal death and will never appear as a biologically functioning life or any living physical structure.

In coherence with visible and invisible infinitism performances, 'Energy' ensures that all aspects of the universal systems function seamlessly and harmoniously together. So, it is appropriate to honor 'Energy' as 'Energy-God' is complete, in whatever 3D ways the community God of the human species has partial dividends.

→From the socio-scientifical viewpoint, first and foremost, the purpose of life is to be competent and be willing to struggle in every moment of survival, even to the molecular level.

This temporal desirement of a molecular level tends to emphasize existence as an independent molecule or a union as a human being in any conducive surroundings, which may call back the memory following many similarities of the physical presentation of one of the principles of natural selection of Darwinism (*that is, the 'struggle for existence'*).

But the same (*that is, worldly observing natural selection*), augmented ways may attempt to match relatively, following the energy levels' works, keeping the rest of temporally related terms of the principles of natural selection, with many locally introduced provisions and rules secondarily connected either for requirements of this bio-nature directly (*like the natural gifts of diverse methods of reproductions and many others*) or obscure indirect purposes for individuals and others relating to the distinctive sections of sociology.

(*By the way, one thing may quote that "Energy" does not require physically introducing reproductive methods since it innately occupies diversifying properties, which determines to solve the sense like reproduction.*)

So, following the temporal sacredness feelings during the struggle for survival, if someone serves with good works in balancing order for saving natural resources that meet a common goal for the comfortable existence of living species and others following balancedness and without overstressing mother nature's resources.

Automatically *(that is limited to 3D discernment*), the good God or helpful or positively inducing Energy comes about to bless an individual and even the whole society in a place of residence.

*{(Indeed, one should not selfishly deny the struggle of motherly caring bio-Globe while she is also consistently struggling following her extended dimensional coverage (*that is, beyond the standard 3D application for human species*) in the Energy -Ocean of the 2nd-energy-belt to acquire some form of inhabitable resource to save and secure her progenies)}.*

→Purpose of life in a more comprehensive aspect.

However, I go more in-depth on the above utilizing some common philosophy, balancing with the contemporary generations, and say, in principle, following the works of mass-energy {*that is, the composite state of unit-value mass-energy or union of energy-elements (ee) & motion element (c)*}, the temporally presented word-sense 'life' has no authentic purpose.

Since, on the point of Energy, 'life' factually is an integral essence of any mass-energy frame (*mc^2 or mc^<2>*) and has broader importance beyond the restrictions that provide the temporal representation for living species in this inhabitable atmosphere.

→The drawing EU74 endeavours to indicate when someone thinks about Energy, the essence of 'life' always exists integrally and has the universal process of appearing and disappearing in the fathomless ocean of Energy, regardless of someone's partial ways of judgment following observable and non-observable matter, material and living or animated and non-animated items.

Only something and someone needs to reserve balance while appearing as a non-living or living status from a most micro cellular state and then evolve as an object or human being enduring or surviving and plying through the place of reasonable energy-atmosphere like this motherly loving bio-Globe.

→Once more, briefly clarify that the interpretation through the word sense 'Energy' is the root cause for everything, including its appearance as a bodily mass with or without discernible as a life form.

Energy somewhat represents the reality of the Divine God by taking superposition in mind of the human species as a belief that God is only the source of life and activities regardless of his (God's) virtual /corporal presence or not, in this world, distance to heaven and everywhere

The drawing EU74 endeavours to indicate when someone thinks about Energy, the essence of 'life' always exists integrally and has the universal process of appearing and disappearing in the fathomless ocean of Energy, regardless of someone's partial ways of judgment following observable and non-observable matter, material and living or animated and non-animated items.

Only something and someone needs to reserve balance while appearing as a non-living or living status from a most micro molecular state and then evolve as an object or human being enduring /surviving and plying through the place of adaptable energy-atmosphere like this motherly loving bio-Globe.

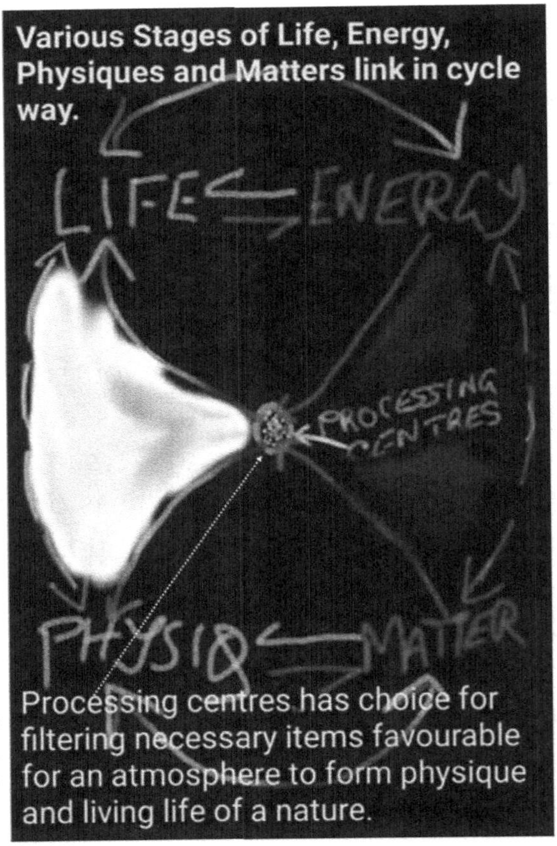

→Life and Liveliness!

Following the quantum world, it may presume that the word sense 'life' is in a consistent procession, irrespective of its temporal significance for a short or long presence as a living being or any micro molecule since the functions of "Motion" or activities of motion-element, determine that fundamental assignment for the roles of life and animated activities of tiny or big living mass.

And 'lively' is its variable significance, discernible either of positive or negative diversions based on the self-bearing constant of an individual or others to prove the activities while existing in a supportive atmosphere.

Simple, like a medical coma man and a fully active man, both occupy life-operating energy to run the physical system. But particularity is the qualitative diverseness, on the point of conscious, semiconscious, or unconscious, when representing liveliness in a physique and performing the purposes of life as a living human status.

►The drawing EU74a indicates a physique of its three stages of particularity on the point of diverseness as the standardness of fully conscious where the body physics is capable of handling surrounding atmospheric physics to some measurable standardness that determines the liveliness of the living status.

►The semiconscious state, where parametrical standardless drop to some lower level, whereas the body physics struggles to check the standard atmospheric physics.

►So, a dead or entirely unconscious condition is when the body physics reaches the most inferior state (*or lowest to a potentialism on the aspect of conceding bodily roles*) from the acceptable liveliness standard and where the molecular union of the body mass increasingly loses balances and incapable of maintaining necessary physics value's functioning and so the molecular level's congruences break down for decomposition.

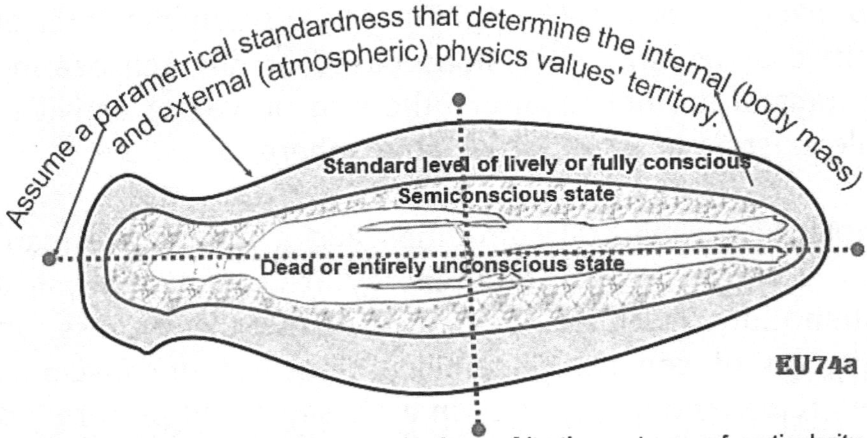

The drawing EU74a indicates a physique of its three stages of particularity on the point of diverseness as the standardness of fully conscious where the body physics is capable of handling surrounding atmospheric physics to some measurable standardness that determines the liveliness of the living status.

▶The semiconscious state, where parametrical standardless drop to some lower level where the body physics struggles to check the standard atmospheric physics.

▶So, a dead or entirely unconscious condition is when the body physics reaches the most inferior state (or lowest to a potentialism on the aspect of conceding bodily roles) from the acceptable liveliness standard and where the molecular union of the body mass increasingly loss balances and incapable of maintaining necessary physics value's functioning, and so the molecular level's congruences break down for decomposition.

→Upon conducting a simple 3D projection, it can be analyzed more with another graph (EU74b), showing alternating waves alongside a straight line representing the advancement of different phases of lively functions from childhood to old age.

The virtuality of 3D discernment determines two ways of deviations, that is, the positive (+) or joyful role, and the negative (-) or depressed state of living status maintaining a covert **constant** of the physique (**'constant'** *indicated in the drawing is as a straight line*).

EU74b

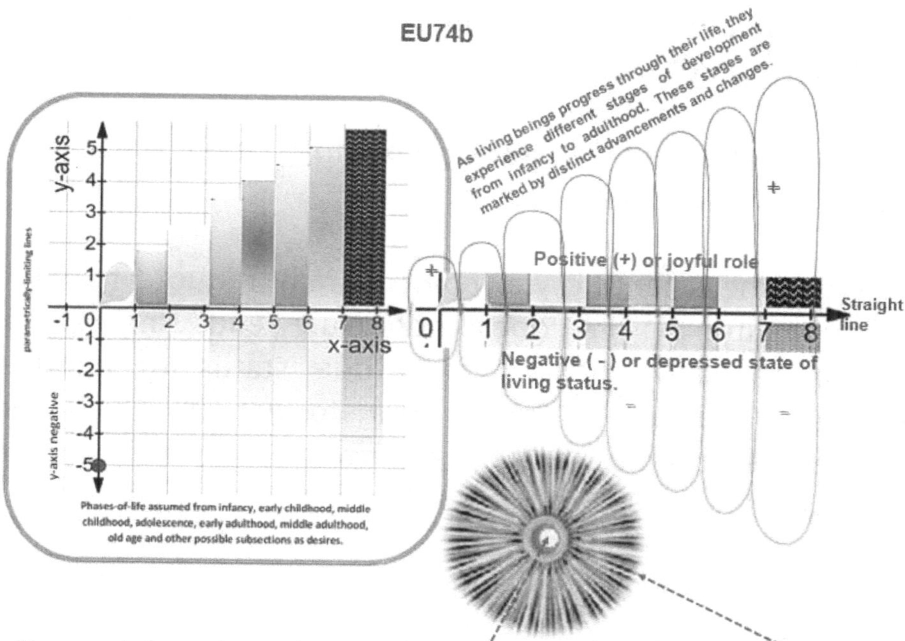

The momentariness of energy from any point of origin is exceedingly faster and multidirectional than a simple 3D, and other supplemented way of measurement, calculation, or determination.

From the standpoint of energy, one thing to remember is that the three-dimensional (3D) ways of judgments always retain one or more controversies from the energy levels' incessant multidimensional covert operations.

Therefore, even someone's 3D way of calculative exhilaration may lead towards disparity since the momentariness of energy from any point of origin is exceedingly faster and multidirectional than a simple 3D and other supplemented way of measurement, calculation, or determination.

(For that, philosophically exemplifies, somewhat like the molecular levels' unaccounted suffering of any wise and unsmart living being from childhood to old age since deficiency pertains to the self-bearing or in-built measurement system that mainly gives importance to

organs instead of each and individual member of the molecular unions.)

xi. From the Energy standpoint, life never ends, and death never occurs.

a. From the scientific viewpoint, it may requote that the functionality of life does not truly end. Instead, it metamorphically changes from one status to another following the law of Energy Transformation, which can neither create nor destroy but can transform from one state to another.

In-depth say, any unit-value living represents some form of mass state energy, irrespective of a corporal appearance as a human being, micro bacteria, or more tiny state of momentarily living role. Following the energy transformation law, upon exhaustion of life-operating energy from a physical appearance, it transmutes to the same or some other form of energy that determines its non-ceasing dignity or eternal status as energy.

{(*Technically, in a body-mass, mass-energy frame, mc^2 or mc^<2>, the alphabet (m) portrays a mass, and (c) embodies the life-operating essence.*)}

So, judiciously, all living beings in this biological Globe are the creations of 'energy-God' (E), and the liveliness of 'energy-God' is everywhere in the 2nd-energy-belt since some form of motion elements (c) integrally persistent, whether discernible through the constitutionally set 3D ways of assessment or aiding with some supplementations.

b. Temporally, death means cessation of the metabolism exercises in a living corporal system, which (that is temporal extinction) no one could so far overcome and make the physical presence unaltering to an eternal lively position.

But someone could overcome by taming the mental system following the science and scientifically invented universal rules of energy transformation concerning the natural ways of engendering the sense of fear while deterring the mindset sway about ceasing the smooth flow of a momentarily enjoyable life someday.

(Moreover, advancement towards the innovations of bioartificial organs using living cells and transplantation or replacement of damaged organs reflects the faith in Energy-God, which presently confines within some community feelings as community God.

Besides, technological improvements sway the mindset of living beings from conservative to wide ranges, which acknowledge the pride towards creator or designer (energy-God) and present them in unifying one form of bio-intelligent species of Mother Nature's biological Globe.)

xii. "Some points about Humanity and Artificial intelligence (AI)."

By the way, in general, the sense of humanity represents the benevolence quality and the standard of living as human beings that meet the contemporary requirements of sociological Goals.

So, in this socio-culturally cultivated bio-globe, human beings occupy distinctive characteristics from many other species that are side by side residing, with or without recognizable sociology and on the aspects of humanity.

→If someone acknowledges the reality of Energy and the universally operational rules and regulations of the 'Energy-God, then the scientific feeling will gradually educate and develop concerning the energy levels practicality in the brain processor of human beings, which will allow overcoming many traditionally following arbitrary practices that may save from unnecessary misery derive in mind and consequentially affect the physique as well the society when trying to implement in the momentarily changing diversity, and so the altered atmosphere and substantial surrounding circumstances.

*{(Something impact like the multicultural society of the human species in this assorted atmospheric bio-Globe, if someone tends to reserve the mindset of traditional brightness and some other intends to jump fast towards modern development before being given training in the biological and mental system's natural (*or molecular levels*) or artificial ways (*or technologically improved method*), how to adapt in the different /isolated environments.*

Also, someone may think of any possibility and consequences of an aquatic mammal and a terrestrial animal intending to swap their dwellings without altering the naturally developed physique or corporal system!

→So, one thing is a matter-of-fact that the Mother Nature bio-Globe is the most reliable and awe-inspiring natural scientist, even though she is a seemingly slower pace than the views, ideas, and thought-provoking concepts derived by the created best intelligent species.)}

(Here, one thing to requote for conceptional clearance, from the viewpoint of Energy, every single thing and all forms of 'acts and effects' directly or indirectly linked throughout the 2nd-energy-belt.)

→Indeed, the essence of Mother Nature's bio-globe, in many ways, circumlocutorily teaches the most intelligent human progenies about proper management of her own reserved and earning energy resources by avoiding destructive attitudes resulting from exaggeration, misuse and likewise practices for self or group interest due to worldly /avaricious attractions and desire to pertinence and dominancy.

(One thing should remember that whatever someone or somebody satisfies, articulately declaring him or them as permanent, but at the energy levels, roles of transformation are most fleeting and consistent, that is, the true philosophy of science.)

This natural education indicates reward points and better chances of beneficial survival to an individual and the group as a species and possibly getting blessed for rebirth as a human being in this bio-Globe or a much better place like heaven's life (*that is, innerly developed socio-religious belief*) to a distant exotic life-supporting energy source, as and since the existing bio-physique make no significance after exhaustion.

→Further, the minds of someone may knock about the life, death, and humanity of modern Artificial intelligence (AI) compared to the existing human generation.

When differentiating between the physical state of a human (*which runs on an exclusive form of bio-grade energy*) and that of an AI (*which runs on other grades of energy resources*), the traditional 3D concept of life and death based on existence will continue to be relevant for acceptance by bio-human beings. It is because the terms "life" and "death" hold significance for living beings based on the energy that powers everyone for a specific period of physically appearing length.

But concerning humanity, which seems exclusive to human beings, natural ways developed following the changing of survival strategy from the ancestor's era while struggling in the harsh atmosphere that carried to successive generations, and so on obeying by the modern age-group comparably comfortable in this easy-going atmosphere.

(Technically, following the energy level's physics values' roles enacted to develop humanity-like things at the molecular level resulted in cooperative works when they unite to perform an organ and then their united effort to save the union as a living being.

Accordingly, someone may have noticed that many individuals are ready to spare any sociologically standard or non-standard of dignity for the kindness of human beings and other living, overcoming individual physical desires, even while struggling in harrowing circumstances.)

So, it is reasonable to assert that the various senses and sense-organs attributed to revered words like sociology, civilization, humanity, and so many expressions were not conceived all at once by the modern or previous generations. Instead, technical influences of the habitable

Mother Nature enacted on disposition and living species that played a significant role in their development.

For the 'AI' essences of humankind programmed following the mix of contemporaneousness of human beings' direct influence, besides indirect of bio-mother nature's influence that carried from generation wise to modern bio-species.

(*It means today's glorified humanity of the intelligent society did not appear suddenly, but it is the resultant outcome that natural ways enacted and allow to improve the human consciousness that cares directly or indirectly to everyone in the society and beyond.*)

→One thing is to remember that the human species is elegant, intelligent, and more flexible, analogously to their invented Artificial intelligence (AI).

Also, eras of natural efforts and blessings shower over the human species from the ancestors, which allowed them to augment many aspects that follow generation to generation with the eternally changing natural atmosphere (Twig + energy-Atmosphere).

Indeed, briefly say there's nothing to worry about 'AI, which will never overcome mother nature's best creations like human beings. But the AI's roles may hail beneficial impact and bliss towards enjoying a cozy life for the inventors of intelligent beings.

Artificial intelligence (AI) may simulate many things for future generations, offering new possibilities for improved descendants. That is similar to how our ancestors persevered in many things while struggling through hardship, which led them to discover new things and ultimately improve the quality of life through new inventions that somehow progressively carry on to subsequent progenies.

So, the concept of 3D ways of discretion as high-calibrated, standard, and worse about humanity is not an individual or community creation of any living species. Instead, universally, everything performs following a chain reaction of the act-and-effect, which is the root cause of in-depth energy level's diversifying functionalities, and that can reveal or be realizable through the developed sensing system of living beings and many other natural-born livings and invented non-living like AI.

(So, it is easy to guess that an 'AI' can perform millions of modern intelligent beings' works altogether with reliable and valuable insights and a much faster pace, analog to an individual intelligence's contributions and 3D ways appraisal.

But the next generations (humans) will be more advanced, analogous to their current invention, like AI, so the advancement for the next generation may not be easily feasible to some modern intelligent beings, which is somewhat similar to the ancestors who never realized their innovative contributions would influence more and more pride to follow-on generations like a modern human being.)

xiii. The notion of religions, its extending preference and energy levels' technical view.

The formal expression of the word religion is a controversial issue and quite complicated since it became a mindset besides its institutionalization as a community feeling involving faith and worship towards supernatural power.

So, sorry to individuals, those who are gentle on that issue, and since the below expression on the subject of socio-culturally related religions has a more comprehensive and in-depth association with energy or Energy-God, none of the terms have any particularity and has any sentimental on the point of community feelings instead, it has broader reality for residing entire human species and humanity in this parentally caring biological planet.

Many may accept that all religions, and likewise philosophical believers, have one principal goal of faith for obtaining blessings from Almighty power /Shakti for oneself or a group of human beings.

But, if someone thinks about the factual source of religion and analyses the ancestor's eras, at least one may guess that they (*ancestors*) were not behind by the faith in some unknown power that provided support and blessings miraculously in many ways while struggling to survive in the surrounding approachable resources in this biological planet.

Commencing from an individual of the small groups of ancestor eras' faith in the miraculous power turned into a beloved Almightiness due to the assistance for many favorable purposes while struggling in the harsh and uncertain environment.

Gradually, with the progression of the human species, that Almighty power somehow for fear, courage, or benevolence, sparked a deepening of individual systems of forebears to form coteries and then gravitate to shape considerable groups before recognizing as the society with some standardness of the essence of culture, believing in the unified blessing strength from almighty nature.

Recurrently, faith in Almighty God influenced and emanated towards enhancing relationships from some individuals to many, for making associations that allowed to obtain accumulated blessings of Almighty God or supernatural potency and to meet a common purpose while consistently struggling for survival.

Accordingly, the substantiality of Almighty God is effectively positioned from an individual forebear system that gravitates to their association representing small groups or entire human species.

Now, even upon the emergence of modern science with the reintroduction of the word sense Energy and its diversifying overt and covert functionalities, the same miraculous power as Energy, Shakti, Almighty God, and locally emanated expressing faith, embodied in the human systems as a loyalty that supports many aspects of life since the essence of struggle became energy levels' innate values for three-dimensionally measuring and consolidate ways of acceptance by mean of the living physique.

Indeed, our ancestors did not know or were interested in the word-sense 'religion' and its modern ways of diversifying methods or community-wise practices and prayers to obtain the blessing of Almighty power for oneself, near and dear ones, and group interest believing three-dimensional heavenly relation of life as Past-life, Present-life, and Future-life.

It is reasonable to say since antecedents were busy struggling and gaining know-how mainly for survival purposes that consequentially enhanced for modernizations with more and more discoveries beyond their (*predecessors*) realization beforehand.

Technically, following the energy levels' roles, the predecessors' discoveries were somehow enacted on human genes. And thereon, direct or indirect (*here, indirect mean energy levels causality*) ways more or less thoroughly, or partly carry to subsequent generations, which consequentially derived changes for coziness from the harshness of struggle straightly, and indirect for survival necessities and physical existence as a living state mass-energy frame or a living body.

So, initially, from the ancestor's eras' there were most infirm or no community feelings concerning religion or religious-like activities compared to successive generations in the easy-going or improved atmospheres.

Following the improvement of the harshness of atmospheres, the roles of integrated organs like the heart, liver, kidney, spleen, lung, etc., besides the limbic or brain system (*configuration of the brain mainly performs as regulators for the molecular union as living physique*), all keep changing following the improvement of survival necessities and invented techniques, and so the standardness from predecessors' status in coordination with survival strategies and surrounding energy levels physics.

Hence, changing the fundamental factors for survival from the ancestors resulted in changing constituent molecules of a physique world like the human being on the aspects of behaviors, cultures, necessities, and many things, including the region and religious faith.

Due to technological enhancements, analogously from the aggressive primeval and medieval community, the advanced eras' complaisant intelligent beings can realize more about the inner meaning of physical requirements and purposes of survival as a bodily appearance in this habitable parentally caring Mother Nature bio-Globe.

Formation of Community solves a group requirement based on geographical availability that generally influences natural ways or by non-natural means to others for an individual or their union as a small or large group for some particular interest.

Indeed, from the temporal /physical standpoint (*here, one thing to remember is that the 'temporal' scenario restricts by the particularity of dimension/s compared to energy level's innumerable multidirectional functionalities*), biasness practices from an individual to a group/s interest may allow one to learn the varieties of cultural requirements and development following the unrealizable energy levels' atmospheric changes.

Also, exaggerating physical desires for anything of an individual or group as a social community may call for or lead to few, excessive, or entire destructions from a pleasing setup. Or the same that is, the destructive attitude may give no hope or a little hope, or 'strugglesome' unknown achievement for new generations.

{Technically, for conceptional clearance, one thing may give a touch and say, following the three-dimensional ways of physical discernment and compared to a small unit-value's mass-energy framed depicted as a molecule, their (molecular) union or big-values mass-energy frame representing a human being, the observable worlds, and universes, the 2nd-energy-belt is immense and non-discernible on the aspects of energy levels multi-dimensional activities and achievements.

*From the standpoint of energy, it is to say, any unit-value, whether defined as a small or massive mass-energy, mass-energy frame of the pre-and-post mature state as discernible or imperceptible objects, all are innerly multi-dimensionally active so that they (mass-energy) can be challenged better and wide-ranging external energy physics, compared to some peripherally strong or weak challenging inflexible dimensionally determinable (*that is, acceptable range*) items or objects.}*

Further, traditionally, the vision of religion involves a complex set of beliefs, values, and practices that shape people's lives, thinking of them as a small or large independent group. And from a technical perspective, religion has multifaceted phenomena that confine various cultural, social, and psychological dimensions.

But at energy levels, it (*religion*) has a deep-seated preference and a powerful force that can influence people's behavior and attitudes for the entire human species to save their appearance and enjoy more securely on this elegant biological planet.

From the above all clarity, will it not be better opinion to think about the unification of all partial and non-partial feelings that diversified sociologically improved intelligent species and come together to save our parentally caring habitable planet for the happy and peaceful life of individuals, as community and even for next biological progenies!!

Following the technical advancement and from the perspective of sociological diversity, will it not be prudency for everyone If we can unify all differential ideas about the almighty power that genomically carries forward from our ancestors' eras to the modern generation and respect as one and only one 'Energy-God' to reach a goal of equality, for the goodness of our existence and new generations to retain highly dignity as gracious intelligent species in this biological planet.

Will it not be everyone's prudent and honest opinion to fulfill the preaching of all heavenly appearance godly preachers from time to time and teach everyone that the Almighty God is one and only one whatsoever, and wherever someone's appearance as intelligent human beings, instead of seemingly be them as independent thinkers and positioned following personal methods of praying while maintaining a common dignity as Energy-God, regardless of the expression of praying following different linguistic words and pronunciations based on some setup practices!!

Also, from a broader perspective of energy, everyone interlinked with Energy-God for this or that purpose while alive or upon physical exhaustion, regardless of individual and group's high pride (+ feeling) or modesty (- feeling) from the perspective of modern sociology in a place of dwelling like this habitable planet.

xiv. A response to a query about the endeavoring to outmaneuver traditional practices is only scrambling of one's brain system.

The brain of a living being is indeed a complex system that consistently works to process information and provide feedback. Its constant analysis enables us to maintain our physical well-being and prepare for the future.

In this regard, I recall the word scramble game. Many people enjoy this game, and some are not. This game is easier to understand than the complex arrangement and partial choice to work of the brain.

The context of the scramble game is mainly to arrange the unscrambled letters to make a meaningful word presented in a dictionary or logic acceptance.

One thing to point out is that in the modern linguistic way of expressing alphabets and words, their unitedly formed sentences are delimited and placed in such that the intelligent system of linguistic groups instructs to accept the arrangement of those alphabets or linguistic elements are in systematic order (*like 26 English letters from A to Z*), whereas to make meaningful words is a random choice from the systematically placed alphabets before making sensible sentences.

So, regardless of whether random or systematic, approval /acceptance of words for common uses is pre-decided due to the majority of choices being acknowledged and taught through media like a dictionary, community broadcasting, and training methods.

If someone thinks a little remote sense that those linguistic practices can be linked with all-natural and non-natural practices and their dominating roles, which everyone is Truly dependent on for all communication purposes, and that practices ease our understanding to realize near and distant while in existence.

→Here may take the opportunity to quote Einstein's famous phrase, "God does not play dice with the universe."

Based on that quote, someone should agree that the natural laws and the natural way to organize our systems are not Truly the ideas of randomness or probability. Instead, Universally, all are linked in a systematic order, in every bit of nook and corner of the observable and non-observable Universe that someone can detect or imagine for authenticating the reality of "**Energy and Quanta Relation**" throughout the 2nd-energy-belt.

→Previously or assumed from medieval eras, the concept of 'Soul' was considered a pure state source of pleasure, heavenly consolation, a witness of past and present works, and a communicator for the future. Since the Soul functions as data storage, while existing, it takes superposition as God to an individual and their group form.

(So technically, the Soul acts as a mediator internally for past, present, and prospective newcomer molecules of a bodily union, which facilitate judging in better ways by the brain processor.)

The significance of the Soul gradually turned to the essence of God in a bodily system that our predecessors /ancestors realized while existence during their struggle with good and bad effects in harsh atmospheres. They genomically forwarded to the following generations as a good God (*like heaven*) and a devil God for unfavorable outcomes.

Even in the modern scientific eras, with the introduction of Energy, the Superposition of God and Soul overtly or covertly remain faith for unknown functional roles to many young generations, regardless of faith in quantum or classical ideology of physics.

Anyhow, in my opinion, God and Energy are performing the same works, regardless of whether known and unknown or inaccessible, through the delimited /anchored sensory organs as bio-intelligent species.

In the bodily system, those molecules are responsible for building living lives, somehow (*or, technically say, due to appropriate application of atmospheric-physics*) systematically organized in natural ways to appear as a whole physique recognized as a bio-intelligent living unit.

However, the problems remain with our inbuilt molecular independent position and their combining efforts for the configuration of distinctive organs, then the group way efforts to operate as a whole Intelligent living unit.

Therefore, the deficiency of delimiting judiciary teams (*that is, inbuilt sensing faculties*) that are most dominant by the capacity and quality of delivery through the sensors for the whole physique.

So, if works of sensory organ/s ceased or collapsed by some means, then the logical judgment/s through the processor, which is set up in the brainbox, turned to illogic even surroundings activities serve as factual to the other intelligent beings.

The same somewhat is possible to match with the scrambled game, where a particular alphabet/s out of 26 words makes a meaningful word while the rest, remain out of choice and in a useless position.

→Now, up to those brief explanations, someone at least can easily guess that the leading role of judiciary systems of the brain depends on delimiting sensing signals capable of accessing and delivering judgments through the brainbox. The remaining interconnected organs and molecules are on standby as a scrambled game, similar to unused alphabets, to accept their consistent releasing commutating signals and judge the reality of many feedback of interception while in an active role.

I think the brain is a natural contribution to almost all living species (*except one or few as per modern findings*) for survival requirements, remarkably performing more or less than the physical standard of the human species. And the brain will go on with exercises to save the union of a bodily system as long as someone is alive.

xv. Here are some new clarifications and collections from earlier books about the Concept of "Energy-God."

Interpretation through the word sense 'Energy' uncovers the reality and the root cause of everything observable and non-observable universes, including its (*Energy*) appearance as a life form in this life-supporting biological Globe.

It is somewhat similar to the presence of the Divine God that has taken the superposition in many human minds as a belief that God is the only source of life and activities regardless of an individual past, the existing situation or present, and the unknown future life while dwelling in this habitable planet.

→From the predecessor eras to modern generations, everyone struggled to acquire some befitting or precious classes of Energy for daily existence and survival necessities, either directly absorbable by some means or consumable indirectly upon materializing as organic and non-organic products or objects that bless them to secure existence at a place of inhabitant.

→The sense of God /Shakti /Almighty and so on the mythology belief, knowingly or unknowingly developed in the bio-intellect systems due to the mysterious /unknown role for something while struggling to exist in native harsh surroundings since from the ancestors' eras.

So, it is reasonable to say that the sense of God is a natural creation instead of the progressing era's linguistically distinguishing for the individual or group's blessing supreme power or divine being.

→However, the science of the contemporary era has advanced much more with discoveries, including the True position of Energy and its requirement for all technical and non-technical purposes throughout the accessible limits.

Also, for living systems, Energy /Shakti became an essential component for the existence and maintenance of the vitality of intelligent species in this bio-nature and even the distant inhabitable nature/s created from some other bio-grade or non-bio grades of life-supporting planets.

So, with the increasing inventions and utilization of natural and celestial resources, the reality of the Energy level's functionalities exposes much more than the antecedents, which enlighten the conservative ways of believing multiple Gods and partial feelings to get high blessings competitively for some desires and satisfactions of individuals and groups.

→As a member of this advanced civilized world and intelligent species in this biological Globe, my respect is high for our predecessors who contributed to the Idea of Shakti /God /Almighty power that directly or indirectly guided towards the improvement of strategies for discovering and overcoming the influence of many tragic mysteries for acquiring the relatively cozy existence technique to survive in this native Globe.

So, in my opinion, for the betterment of the contemporary, the next progenies, and other existing creatures, everyone needs to be thoughtful much more in advance by overcoming partialism, feelings of dominancy, and reign as an individual or their group's ideology, considering the supreme functionalities of Energy and the Energy-God over the dwelling planet.

→From the perspective of Energy, every representing individual, their united groups, and all other living life have a homologous relationship with the 'Energy-God' in all aspects of existence regardless of the place of living and physical dignity /identity in this parentally caring biological planet.

→One thing is to remember that every individual large physique, or physical world like a human being, represents a union of molecules, virtually regulated by the in-build delimiting sensing faculties, which mainly solve some purposes, whether enjoying the exquisite situation or suffering in a severe place of living and uneven circumstances.

However, technically, at the molecular levels of the physique, every molecule is consistently busy with stemming and challenging internal body physics and externally plying atmospheric physics.

Therefore, in all cases of internal safety of an alive physique, overall body mass, mass-energy frame, or individually representing Energy-God, always covertly tends to save and secure from reasonable unfortunate or unfriendly external atmospheric application, whatsoever someone claims differently following individual decisive behavior based on 3D (*three-dimensional*) judication.

→So, from the above, one of the fundamental points easily comprehensible that three-dimensional ways respected Energy-God always tends to perform for the goodness of an individual, provided any of the constituent components, including the temporal purposes setup sensing faculties and brain processor of the physique does not gets overindulge for many (*here 'many' is, either or both the materialistic and celestial leverage, means worldly and transcendental attraction)* reasons that led to loss the overall control of body management due to the influence of external unfriendly or devil physics.

Besides, as a consequence of the above simple physically engendering non-natural cause (*mean, something like obscure /hidden wrong works of an individual or group satisfaction*), many standard natural factors seemingly function as devil roles even to a healthy living system welcoming some form of early misery or casualty.

So, knowingly or unknowingly, invitation of the unusual negative or devil physics from the usual positive or angelic internal body physics, now and then witnessed by everyone from an individual childhood to the end of prevailing periods as living, which effectively allows the bodily system to experience all the events (*that is, high and low of some reasonable standard*) of joy and sorrow while in presence as a living state.

(By the way, one thing is to remember, at the energy level, one's internal system includes the brain system, always witnessing even if someone is confident that nothing is all around and no-witnessers for an individual's wrong-doing. Technically, it is due to the natural way projected three-dimensional layout of a living system that restricts an individual in many aspects.)

xvi. Further, herewith are accumulations of some points mentioned in the earlier books for a better acquaintance of the faith in God from independent individuals, their group, and the entire biological intelligent species, besides the scientific ways of exploring 'Energy' and its practical presentations for all purposes throughout the Universe.

I think everyone consciously or instinctively believes in the works of God and experiences gained by our predecessors, and many functional essences of God that genomically or other remote energy levels mean, which carried forward as a natural gift to follow-on generations, even to the contemporary intelligent beings. And, even those who are in orthodox faith in the practicality of modern physical Science (*that is, perceptible limits*) and physically represent them entirely atheist.

→It may quote, expression as a 'practical situation' or the sense of observable practicality and the physical functionalism restricted by the works of the internal system as a bio-intelligent being, where in-built sensory faculties (*like eye, ear, nose, skin, etc.*,) play the most significant roles and generally perform the accessible limits directly (*that is, straightly through the physique*) or indirectly (*indirectly means, supplementing methods*), whereas repressing many molecular functionalities and their daily requirements while in existence as a living union and lively status.

→In the modern scientific era, upon the exploration of the comprehensiveness of 'Energy' and in a formulaic way presentation, E = mc^2, I judge that the equation 'E = mc^2' is the best gift for biologically materialized modern intelligent beings and the progressing follow-on biological generations.

Also, the equation can be utilized to characterize the most-micro invisible living state mass-energy frame and the molecules that participate as a union of body mass of living beings.

→ (*The drawing EU75 indicates a simple model for ease of understanding that 'Energy' can be utilized in different ways defined in many places of earlier books.)*

The drawing EU75 indicates that Energy (E) is a discrete amount of unit-value mass-energy while occupying some shape of 'mass state life-orienting bio-grade-energy' with constitutionally sets some levels of energy elements or working essences, including the soul or motion-deriving motion element (c) that overall characterize /denote as a compressed sack of life-orienting bio-grade Energy or a mass-energy frame represents a living being.

→One thing to remember is that the gift from our predecessors is something like 'God' and the presentation through modern science like 'Energy' both the word expressions retain equivalence relation and are responsible for all forms of energy products, objects or creations, including diverse states of matter, material, materialistic items besides the bio-living beings and their organizing socio-logically related bad and good derivatives.

Those sociological products like Socio-cultural, Socio-political, Socio-Ideological, Socio-scientific, Socio-economic, etc., of biological beings, whatsoever someone's discretions while lively but preferably choices, either of one or all as a non-believer, or believer of God and Energy as a common one, but the 'essence of Energy-God' never breakdown.

However, one thing is standard to all about desirement and prayer, to get some form of satisfaction either of good (*positive or favor of something desire*) or bad (*negative from the standardness of common interest*) purposes or attempt to neutralize internal turbulence (*that derive in mind*) due to some known or unknown reasoning.

Some state of a compressed sack of life-orienting bio-grade energy or a mass-energy frame represents a living being.

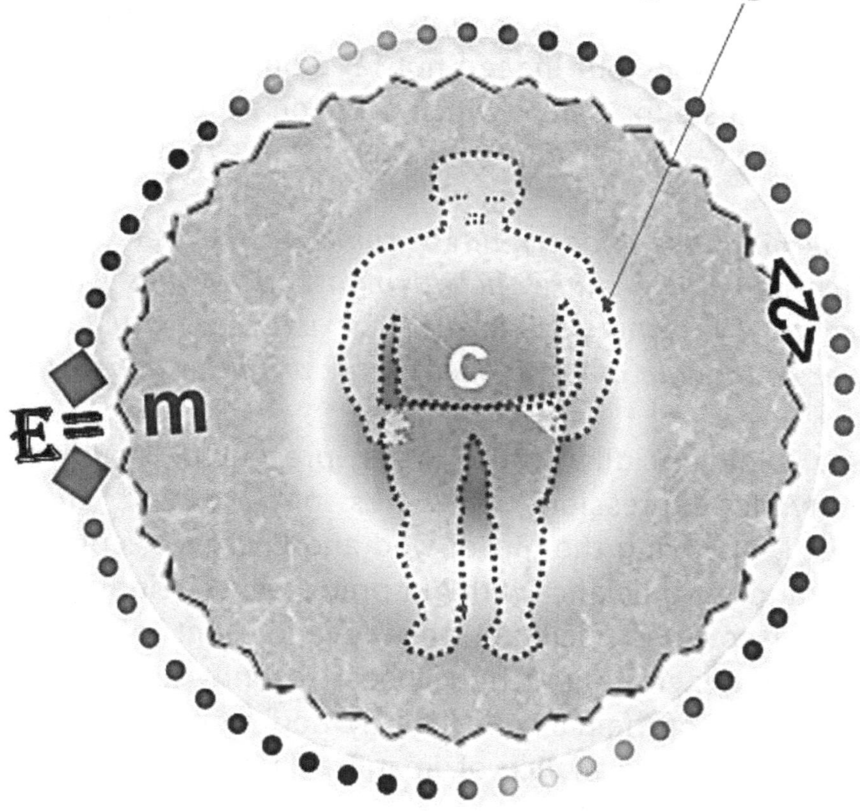

The essence of Energy (E) is everywhere within the frame <2> of mass-energy (m) and motion-deriving energy component (c).

EU75

Likewise, the sociologically implying expression as God, Energy, or Energy-God can apply to all desired values and undesired purposes that generally grow into the mind of naturally structured bio-intellect systems and even to some standards of that desire or undesired (*means, which mainly acknowledge for living beings*) is somewhat relevant for the Artificial intelligence (*AI*) being those are seemingly (*here, concerning the sense of seemingly for AI is already defined to understand, in this chapter and chapter 14*) direct human creations but indeed not far from the technique of living beings expression, which may coexist along with the next improved generations for each other's requirements.

(*Something like modern-day's some of the pet animals, which can fairly share their sympathetic values with pet parents. The same standard of coexistence attitude could never be thought of or realized by the primitive generation.*)

→So, avoiding dilemmatic ways presented synonymic, antonymic expressions of works of God and Energy. Instead, justifying the gift of God and Energy, keeping the admiration and balance with our predecessors' findings and their discoveries facilitate the advancement of modern Scientific generations. I find it best to unify and honestly dignify both of them as one 'Energy-God' since it is the root cause of all performance of imaginable and unimaginable works of the Universe.

→One thing to quote about the future of bio-species, irrespective of being a believer in God, Energy, and 'Science' all of them have a common ideology concerning death and birth based on the remote theology of reincarnation to the next life upon the occurrence of death.

Since the sense of reincarnation naively imbued in the living system following the belief of predecessors, and then the same sense as 'reincarnation' for the next generation that hints at an approachable logic following the rules of energy transformation (*i.e., changing the 'state of energy'*) from one form to another, which may mean, reincarnation of physical status possibly happen to a biological or other form of energy-mass that can behave an angelic, near angelic or obstinate characteristics upon revealing as a physical appearance.

→By the by, considering the improvement of Scientific discoveries, someone may imagine that one day the biological intelligent progenies will be able to overcome the undesired thought of 'death' and what it means, a state of changing lively appearance, and be alive eternally in a place and condition of happiness or exquisite life forever.

Indeed, imagination is a short satisfaction to our bio-intelligent species upon getting a prideful position, cooling down some of the self-built differential values' functionalities.

{(*Besides, briefly quote, at the energy level, differential values' functionality never allows something as a physical state to stick at a point of inexhaustible constancy.*)

(*Also, remember, at energy levels' momentariness is so fast and immense that no one can practically stop the outcome with differences. Unless and until some amount of energy organizes to form a mass-energy, mass-energy frame or object to get an independent measurable understanding of comparable strategy.*)}

→Furthermore, the Universe is not like an independent physical status representing a simple energy-world, having individuality compositae with a little discrete amount of mass-energy, the acting alive body mass.

Besides, the universal 'Energy-God' takes care of the infinite limits of the immense Universe, including the 2nd-energy-belt.

So, it is reasonable to say that diversities are Energy's creations when satisfaction is the gift for amusement while the creatures reside in a place and Nativity or travel through the facilitative surroundings.

→Therefore, the mysterious works of 'Energy-God' and generations after generations from the precursor's discoveries prudently carried over by employing the outstanding technique of Energy levels' functionalities in coherence with the progression of the living form mass-energy or in the form of organic level's living beings' physical systems' mind and morally developing judicial court in coordination with the altering inhabiting atmosphere in this parentally caring habitable Globe.

→Accordingly, it is convincible that the expression through the word sense 'Energy' has scientific era's logicality that represents the Superposition like 'God' in the minds of antecedents and genomically carried over the follow-on generations in the congruence of the dominated mysterious surrounding, which providing time-bound satisfactions while appearing in this habitable biologically fertile Globe.

→Further to quote, at energy levels, the practice of evolution, devolution, and regression with the diverse role of the differential coefficient is ever persistent in this 2nd-energy-belt.

Accordingly, consistent roles about the development and discoveries of mysteries from the past, followed by the next generations where they find something altering impressions, which replaced the accepted faith in some other states with different expressions such as 'God' or some powerful source like Energy.

→However, in the modern scientific era, the role of 'Energy' became more prominent. Hence, its (*Energy*) authenticity became acceptable as a 'sensible reality' to many intelligent beings than the expression and faith in 'God' that traditionally carried following some heavenly faiths.

→Therefore, the unifying word like Energy-God is my best opinion for faith and prayer in this technologically enhanced civilized generation.

xvii. Herewith are some reasonable characteristics of Energy-God, which accumulated from earlier books, Volumes 1 & 2 of "Mythology, Philosophy, and Science, and possibly adjusted as appropriate for simplification on the issue of Energy-God.

►The Fundamental Status of the Energy-God.

Fundamentally,

The Supreme–Universe
= E
= Energy
= Energy-God

= **True position of Pure-energy at a state-of-Absoluteness**, or say, as **True-Singularity** with all compositions (*that is, Energy-Elements and its inherent working property Elemental-values*) of the Universe that includes the living and non-living products procured.

→Energies in the 2nd-energy-belt is one of the residual derivatives upon distribution in the 1st-energy-belt and the 3rd-energy-belt while in the progression of maturity from juvenescent of the Universe.

→From the perspective of Energy, in the 2nd-energy-belt, nothing may Truly be justified as 'Singular and Singularity' since diversity primarily ruled throughout the coverage of this energy-belt in the matured Universe.

→The observable, non-observable, or objectually characterized word sense 'Singular and Singularity' is at their independent non-differentiable Zero-status-stable position, which has virtual status and the Superlative art of Energy-God.

(*Regarding the sense of the word "Singular and Singularity" for implementation at energy level is elaborated in this and earlier books.*)

▶Locale-derived esteemed Gods and Goddesses of any living nature are the spinoffs of Energy-God.

From the viewpoint of Energy, all small and large groups of religious beliefs and the esteemed Gods and Goddesses of bio-intelligence have taken positions based on surrounding support of socio-culturally, socio-scientific, socio-political, and socio-economical security.

Hence, all tools that regulate society and work in harmony with nature are resources for living, which also accords with sociologically distinguishing and admiring Gods.

Mostly, locale Gods and Goddess worked on balancing orders in concurrence with Energy-God, representing for bio grades energy sources of this adaptive living bio-nature and many other life-supporting natures, so the living creatures get entertained (*in the form of joy or positive and sorrowful or negative*) based on their state of physical appearances.

But sometimes, it seems the diverseness of personal Gods and group beliefs, irrespective of conscious or unconscious positive and negative values' ruckuses and destructions are dilemmatic inclusion by some individuals, following overindulgence temperaments due to dominating attitude for materialistically generated gravitate attraction, which may remain aloof from the principles and immediate actions for balancedness by the universally regulating Energy-God, so the possibilities of unchecked suffering in many aspects.

►**All creations of "Energy-God" are transitional travelers in any of living natures and observable and non-observable universes**.

In the 2nd-energy-belt, no such observable and non-observable items may denote or declare as Truly ever-persistent except the 'Energy-God.

Materialization of all the noticeable and non-observable items persists on some Time-bound natural or non-natural configurations following the rules of Energy transformation.
So, the formation and disappearance of objects are persistent, maintaining the significance of Energy-God unchanged in all aspects of the 2nd-energy-belt.

Besides, expression by natural and non-natural means, or through the different words and configurations in sentence form to some sensible meanings, either bearing the properties of negative, positive, null /zero, etc., are the direct or derivative creations of Energy-God and have significant value irrespective of acceptance for some purpose, or total rejection by the intellects in this and alien natures of the 2nd-energy-belt.

►In this 2nd-energy-belt, Energy-God is entirely balanced.
Universally, 'Energy-God' is always balanced. Besides, Energy-God supervises to protect and reserve her observable and non-observable creations perennially throughout the coverage of the 2nd-energy-belt.

But, exemplifying with independently superior creations like intellect beings of this bio-nature and distant alien intelligent natures, some of the creatures at some point of known or unknown purposes 'more or less' lose out the self-governing control for some reasons and derive towards exaggeration for individual interest or a group species, vainly to get pride of some equivalent authority as superlative power like the local God or the supreme Energy-God.

That state of physical high pride may lead an individual creature or their group towards unrealizable loss from the standardness of balance at a place of appearance or dwelling.

Whatever the cause of losing the standardness of balance by the local creatures or the Gods or Goddess, the Energy-God never gets affected since it follows vast and multi-dimensional roles that allow him (*Energy-God*) to retain eternity in all aspects of dimensional discretions.

►Fundamentally, sociology and sociology-related derivatives directly or indirectly follow the locale or naturale influences in concurrence with the persistent roles of universally supervising Energy-God.

From the viewpoint of energy, all small and large groups of religious beliefs and the esteemed Gods and Goddesses of bio-intelligent beings have taken positions based on surrounding support of socio-culturally, socio-scientific, socio-political, and socio-economical security.

Hence, all those regulative tools derived through sociological means enable performing, in coordination with the natural way provided adaptive resources for subsistence.

Whereas 'Energy-God' is covertly functional, maintaining the universal balancedness status without losing its eternal dignity and persistent supervision throughout the 2nd-energy-belt.

Besides, the earthly or temporal ways emanated Gods and Goddesses capable of playing the roles in specific areas, instead of the immenseness of celestially existing Energy-God with extreme-fast and True impartial functionalities.

In this biological earth, sociologically regulative distinguishing (*something like the sense of 'positive, good or moral supportive,' and 'negative, wrong or immoral,' and 'neutral or equitable' featuring*) Gods and Goddess are limited to particular resources as matters, materials, and materialistic products, or togetherly all those may say as objects, and their diverse derivatives that are seemingly responsible for deriving construction and destruction following the rules of energy transformation or transform to Energy-God relinquishing all identity or dignity as the objects.

Usually, this transformation practice from objects to 'Energy-God' indirectly secures the reasonable state of balancedness and orderliness function of life-supporting energy resources and other necessary derivatives for living creatures, 'unless otherwise' invites annihilation of the habitable Globe for some reason.

► **From the standpoint of energy or Energy-God, and on constructional aspects such as unit-value mass-energy living frame/s and the energy levels' energy elements or working essences are extremely flexible than the representing unit-value individual living or their group as species.**

Indeed, biologically positioned individual livings or their group as species are constituted or ingrained with 'bio-energy genes' or hereditarily obtained specific class of life-orienting energy resources from the Energy-God.

However, nourishing with locale laws (*i.e., local but natural way management by providing rules and regulations*) by the parentally caring biological globe frequently modify as per the conditionality of earnings of life-supporting sources and surrounding situations of this habitable planet in the energy-Ocean.

Therefore, its' (*functionalities of Energy-God*) consequences act as derivatives from the observable sense of objectively materialized Gods and Goddesses, presented as individualism based on personal or group-loving choices.

Whatsoever, they (*locally esteemed Gods and Goddesses*) settle and share the pride of Energy-God, which is available with Mother Nature bio-Globe.

►Locale engendering faith and prayer on divinities somehow soft-soap the Energy-God regardless of positive or negative of individual or groups' desirements.

Accordingly, it may be mentioned that the methods of sociologically bearing faith in many mythological Gods and Goddesses with distinctive dignities and functional impressions are local creations involving local deities and supplicants, directly or indirectly connected with Energy-God.

However, all the natural way derived Gods and Goddesses have the truthful form of solo origin, the Energy-God, an all-mighty foundation throughout the Universe, instead of diametrically limiting distinctive natural Gods of native and distance adaptable natures and living creatures.

►Why does the physique of an individual need proper managerial control to enjoy a happy and peaceful existence?

It quotes that proper managerial control is essential, wherever materialized as a mono or group-designed living state following the survival strategy to preserve the bodily appearance in the ever-evaluating and devaluating practices of existing nature due to the universal act of Energy-God.

Otherwise (*that is, improper management of bodily system*), the natural way devil God comes into effect the system, compelling to disturb /mutate regular works and rules of Energy-God that may call for early calamities before one's realization and God's assistance comes into effect.

Besides, proper management of bodily systems brings contentment, avoiding internally or externally engendering exaggerated attracting force that only gives a short burst of pleasure and happiness and may invite casualty to mind for a long-term perspective,

► "Energy-God" and its multitalented roles.

Considering the sense of extraordinariness or multi-talent works, I feel pleasing to say that only the "Energy-God" and its versatility eternally perform as a multitalented individual, wherever and whatever it represents with volumetric or non-volumetric honors in this 2nd-energy-belt.

→Or philosophically, simple ways concede that Energy-God is 'all in all' with unending ranges of talented activities irrespective of its (*Energy-God*) presence as the biologically formulated superbeing, their culturally and sociologically admiring works and acknowledging diversified Gods as an angelic or devilish, counting on specific desired values' satisfaction or dissatisfaction upon finding an outcome for physical requirements.

►Shape of Energy-God.

Based on the significance of materialistic worlds with specific volumetric shapes and sizes, someone's mind may knock the question about the 'Energy-God' on the aspects of its structural honor.

→In this regard, it may attempt to cross 'over and above' the limits of objectual activities (*means, function related to observable and non-observable objects*) and enter the realms of the fantasy of non-structural energy universe.

→Generally, in this biological earth, socio-religiously depicting God and Goddess are at their positive, negative, and neutral features with beautiful, ugly, shape, no-shape, or sense as 'no existence of such things' to the ranges of someone's individual and group belief.

→So, the limitation of that intelligent query (*that is, the shape of Energy-God*) may extend more and beyond some standard structural ideas, and Imagine that God has no volumetric significance with peripherally sets sensory faculties, like eyes, ears, nose, tongue, skin, etc., besides many interiorly formed complex sensing organs that gave dignity as super-livings in this bio-nature, specially positioned in this biological nature to struggle and survive in the contingent of atmospheres *(that is, Twig + energy-Atmosphere*).

→Instead, if someone can judge on logical values depictions and consider that all those diversifying sensory faculties mainly function as surviving tools, gifted in natural ways due to effective yields of Energy-God.

Then, conclusively, it is rational to say that the Energy God does not require diversifying shape, size, structure, etc., which are particular ways defined within the limitations presented by the domains of socio-cultural, socio-religious, socio-scientific, and likewise non-celestial, non-spiritual materialistic distinction.

(*Or, based on someone's celestial or heavenly faith, it may simplify and say that 'God or Goddess' does not require audio or visual organs that structurally and distinctively facilitate many living and living beings to listen and look at an individual or group's heartfelt prayers.*)

→Energy truly represents the position of God and Goddesses, irrespective of any objectually defined volumetric shape, size, and particularity of feature /property.

And when the Energy-God represents a living mass-energy frame, its constitutionally ingrained most micro-unit energy elements with solo positioning 'motion-element' function as unalloyed 'soul' for the living unit.

→Based on physical, observable nature or objectual logical sense, the reality of the 'Energy-God' is restricted by the limits of perception that usually estimates some discernible dimensions.

So, the nothingness of the mind or perceptible sense considers or judges as there is nothing and even no impression of energy levels' consistent diversifying activities.

Also, according to the standard features of Energy, it (*Energy-God*) has its universal dignity of lively activities whether observable and non-observable objects acquire diversity with differences, or Zero-status-stableness with the logic of neutrally, or agitative and biasedly active living and non-living structures.

→In this bio-nature, many celestial - preachers, every so often embodied as super-living life, guide the reality of One God, i.e., 'only One God' for all, that is the Energy-God, who gives all energetic livelihood of presence as volumetric contemporaneous of categorizing living and non-living structures.

→So, short existence with diverse shapes and sizes as living and non-living energy creatures or energy products determines the presence of God in different ways with or without shape, regardless of socio-culturally, socio-scientific, socio-economic, socio-religious, and other ways of modern civilized intelligence species.

xviii. Further, I wish to remember some points from the Volume II of the book "Mythology, Philosophy and Science" as a prayer for the goodness of all creations on this biological planet, formed from the most honored 'Energy-God' whose incessant blessing gave the existency of living species and secured their presence in the consortium of conducive habitable atmosphere.

As one of the most intelligent creations of Energy-God, it is everyone's responsibility to preserve the dignity and habitability of bio-nature by avoiding all forms of commercial exploitation and exaggeration about the use of resources. It includes protecting living creatures and non-living products and maintaining a balanced order of conduciveness atmosphere for survival.

It is crucial to everyone, as the highest-ranking intelligent species on this biologically flourished planet, to care for all life-sustaining holdings by avoiding all exaggerated desires from materialistic gravity. Such desires can lead to the downfall of labouriously evolved human society, all species, and even the attractive habitable Globe.

Besides, sociologically characterized all positively and negatively featuring expressions like joy and sorrow, good and bad, heaven and hell, and so on may link to the part of physical pursuits that momentarily appear and disappear, giving some effects at the molecular levels of the bio-physical system to all living beings.

→Overall, I believe that the Energy God has accomplished his best effort to construct human beings from a specific grade of energy, unifying billions of lively molecules. Fortunately, the living beings' system is equipped with an exclusive type of regulatory system through the brain to maintain the unity of the physique.

►Lastly, I can poetically talk about the eternal relationship with the Energy -God of the 2nd-energy-belt and the countless creations, including objects and objectual-products, and the Energy-God is only the 'self-regulating' almighty celestial God who eternally exists.

I am the Energy-God!!
I am the "cause and effect" for everything!!

I do not have any beginning!
I do not have any end!!

Even I do not have an ancestor like you!
But I am the progeny of Supreme - Universe!!

Someone may think me,
I am only the cause of all good and evil effects!

But I am not!!

I am the energy,
and I am the "God" for all works!!

You may find me neutral,
You may find me partial!

But I have no choice but to discriminate!!

You may find me only around you,
you may find me everywhere!

But I am the Energy God, and I am Universal!!

When you think of me differently,
you get my diverse acts!
When I act for something,
you get my effects as good-God or bad-God!!

I am the Energy, and I am the God!!

When you find my disparity,
you get my differential values works!
But I am the good for all as Energy-God!!

My work can traverse through heaven and hell!
And can traverse through
any micro-molecule of the Universe!!

When I work with differential value,
you get the consequential "effects"!
I am the Energy God!
Master of differential works!!

My existence is everywhere!
So, my "acts and effects" are everywhere!!

You may find me nearer to your sensible limits!
You may find me beyond the frame of fantasy!!

I am eternal and will remain eternal!
Even the appearance of my transitional creations play diverse roles as an object or objectual product!!

I am Energy-God!
I am the soul of all creations regardless of judgment as lively or lifeless!!!

Note: - I tried to explain thoroughly to my best with the possible practical and assessable philosophical sense in the series of the book "Energy and The Complete Universe" even though I feel sorry for them and their uncomfortableness to comprehend the fundamental energy levels works of the Universe and the extreme infinite diversifying roles, exclusively for the energies in the 2nd-energy-belt.

(One thing to quote is that the True form of Energy is unimaginably flexible than some distinguishing state of matter, which most people are familiar with.)

These concepts of the books may seem unrelated to some people following the traditionality of choices, which has developed through genetic and natural influences carried forward from generation after generation from prehistoric eras.

"I did not have any bad or good intentions while writing the books that align with the current beliefs or values of contemporary consciousness of any individual human being or the sociologically involved group that focuses on the routinely set social stratum, echelon, religion, secularization, law, sexuality, gender, deviance, or any other related aspect."

Furthermore, I would like to state that all the explanations provided in the books more or less followed in-depth reality.

Explanations in the books, frequently accompanied by some sensible quotes and clarifications aiming to convince the preaches of saintly appearing intelligent beings from mythological eras, and then the improved scientific eras technicality that covers all the representable formulas of the Universe, which presented in a simple equation form E = mc^2 to define that Energy and it's all forms of realizable or incomprehensive Energy products like objects and it's any form of derivatives are interchangeable, and eternally maintained equivalent relations.

So, I hope, if not this generation, the future generations will be able to realize the practicality of Energy-God and its universal functions.

The functional entities of Energy-God govern and regulate everything from the smallest particles to the most complex visible and non-visible objects. Also, Energy-God is always impartial and does not distinguish between living and non-living items in the ocean of Energy of the 2nd-energy-belt.

Conclusive descriptions of the book.

The series "Energy and The Complete Universe. Part 4" synthesizes advanced ideas on consciousness, energy, gravity, and their interconnectedness, offering a comprehensive overview of the book's exploration into the nature of life, the Universe, and human existence.

►It explores the multifaceted nature of consciousness in living and non-living intelligent systems and its interactions with various energies, particularly those in the suitable nature of this matured 2nd-Energy-Belt. It delves into how consciousness can be lost or altered, examining the technical aspects that govern its existence and influence. The validity of sensory perception, such as reality, locality, and entanglement, plays a crucial role in understanding consciousness, particularly in multi-dimensional spaces. Human consciousness is restricted by the healthiness of brain processors and processing capabilities, constrained by three-dimensional physical computation, which mainly pursues comparable strategies.

Further, it may elucidate a little about the physically derived 'consciousness' that humans typically experience standardly emerges from complex neural processing and is constrained by their three-dimensional, physical worlds. This form of consciousness is highly dependent on biological factors such as the brain's ability to interpret and process sensory input.

However, when endeavoring to shift focus to the concept of True energy form, particularly in the vast energy -Ocean of the 2nd-energy-belt, encounter a more fundamental form of consciousness, not bound by physical limitations.

Technically, it acknowledges a naturally evolving physical system, 'consciousness' appears to be a product of biochemical interactions, which are subject to decay, alteration, or even complete cessation when the energy system (the body) deteriorates. It is a byproduct of the containing and energetic limitations of the 2nd-energy-belt, especially to a living nature like the biologically developed habitable Globe, which far corresponds to rudimentary forms of energy and consciousness, mainly bound to biological or material decomposition that is available in a place of the inhabitable zone.

However, the energies in the 2nd-energy-belt represent a more maturely active and more evolutive in the actual Universe, where the original consciousness remains intact and pristine even upon maturity.

So, consciousness is no longer tied to physical processes alone but operates holistically and self-sustaining, allowing for eternal regulation and interaction of reality. For the matured 2nd-energy-belt, consciousness is recognizable following the regulative functions of energy, which is more enduring and resilient than any physical state of living and non-living unit.

It is possibly known to everyone that, unlike the unstable and transient consciousness connected to physical systems, the consciousness associated with the energies of the 2nd-energy-belt operates differently. Consciousness in the 2nd-energy-belt remains intact upon reaching maturity, a state where it becomes self-aware and eternally functional. This higher state of consciousness is immune from degradation because it stands on a stable, non-physical foundation. Here, the energy dynamics are not regulated by entropy, decay, or other phenomena from the mundaneness of the physical universes but by eternal principles of balance and harmony.

For living beings, the concept of consciousness has profound implications. So, if the expression 'consciousness' derives from physical forms, which are subject to decay and limitation due to the constraints of the body and brain, then its higher-dimensional counterpart in the vast energy - Ocean offers the potential for continued existence and regulation.

This understanding suggests that even though physical consciousness may fade. But the true essence of consciousness, when aligned with the matured Universe's energy, continues to exist and function.

This eternal regulation is akin to a cosmic oversight mechanism, where the original consciousness continues to influence and shape reality. It operates in a dimension beyond sensory perception and material existence. But persist in maintaining order and coherence in the Universe, ensuring that all systems, whether living or non-living, remain interconnected through this unalterable form of energy-level consciousness.

Therefore, 'living beings,' although constrained by their physical forms, are intrinsically linked to this higher, eternal state of consciousness at energy levels. While the physical brain's processors restrict the perception and interaction with reality to three dimensions, the matured consciousness in the 2nd-energy-belt offers a deeper timeless connection to the Universe's true nature. This energy level's matured consciousness governs and sustains everything at an energetic level far beyond what is observable in the physical world, indicating that life and consciousness do not end but transition into higher states of existence governed by eternal energy principles.

►In the context of 'Gravity,' this book provides a deeper understanding of gravity, which plays a crucial role on Earth and throughout the matured 2nd-energy-belt, from its molecular-level effects on biological systems to its broader cosmic functions. It discusses the technical aspects of gravity in physics, including gravitational fields, force fields, and orbital dynamics, and how these principles apply to energy fields and mass-energy frames. This book discusses the daily application of gravity, particularly in human intelligence and activities like football, where gravity impacts physical movements and even spectators.

►This part of the book series, technically and philosophically, defines the objectually standard expression 'singular and singularity' with a focus on its possible augmented application to energy for the Supreme - Universe and the Universe beyond the observable universes.

►Everyone recognizes the significance of 0/Zero, a fundamental concept in mathematics. Furthermore, the standard concept '0' has grander implications for understanding how it functions at energy levels, like the 'Supreme - Universe,' the actual Universe. Also, for the observable Universes and the energy - Universes, particularly in the 2nd-energy-belt. This zero /0 or 0-like (something, like 'nothingness') sense plays a vital role in understanding the founding position before forming any energy structures (energy structures may mean 'mass-energy frame') and fields.

►Like the other parts of the book series, this book also discusses some points on energy, motion, and motion-related concepts. It highlights the interaction between motion (motion derived due to the roles of the most fundamental type of energy element or energy component) and other 'energy elements or energy components' and emphasizes how a unit-value 'mass-energy' crusades its fundamental dynamical performance in all observable and non-observable systems. The book analytically explores the intricate relationship between motion elements and other energy components intrinsically linked to broader universal energies.

►One should not be confounded by the overall outcomes of classical objects or microparticles when attempting to analyze something at the most fundamental energy level. Remember, "energy" is Truly the complete essence of everything. Due to the extraordinarily prompt componential nature, comprising in-built energy elements and their integrated values or working essences, in any assumable unit-value mass energy, even the energies in the vast 'energy ocean' of the 2nd-energy-belt, play the most momentarily dynamic roles. As a result, primarily, energy becomes a self-dependent and self-regulating authority whenever and wherever it manifests as a classical object or microparticle of distinctive or indistinctive substance.

►Following the concepts of the earlier book series, this part of the book revisited and clarified some points on constructional aspects of the Universe, tying together essential principles and ideas on how energy and matter interact to shape the cosmos.

This part of the book series explores the paradoxes of life and death from a quantum and energy perspective, challenging traditional definitions of being "alive" or "dead." It addresses common misinterpretations of terms like "dead" and "decay," demonstrating how these concepts may not fully apply in the context of energy.

►Additionally, it hints at the spiritual vision of Nirvana and its potential benefits at the energy level, particularly about the painful emotional instability in living human beings.

►Furthermore, it discusses the concept of 'free will' and its limited application in bio-intelligent systems as a critical topic. It explores how beings make choices based on their physically evolving consciousness and the continuous influence of their internal and external energy levels' operations, whether resulting in good, flawed, or irritating decisions in various situations.

►It discussed the importance of civilization for humanity's development, which gradually enhanced from beginning to approaches generation after generation following the surrounding energy level availability and reasonability of mother nature for adaptation in a place of habitation.

This book explores various miscellaneous concepts related to naturally evolved survival mechanisms and their impact on physical systems.

►From an energy perspective, it also analyzes how the concept of "life" and its significance extends beyond the traditional human-centric view, embracing other forms of existence. The standard belief in the binary nature of life and death does not fully capture the true meaning of these terms. Instead, from the standpoint of energy, a core principle emerges, 'life never truly ends, and death never actually occurs,' only the transformation of physical forms into energy and back again.

►This part of the book series also explores the relationship between humanity and artificial intelligence (AI). While AI can demonstrate high-capacity performance based on its design and intended functions, it is limited when competing beyond the individual level, especially as part of a larger living species. In the broader context of a habitable planet, the naturally evolved human species remains more adaptive and versatile in many ways. It is because the Earth's biosphere, shaped by Mother Nature, continually nurtures life and introduces subtle energy-level changes to all its progeny, including biologically evolved living beings.

Based on the above explanation, this book series pragmatically suggests the possibility of extraterrestrial life forms that have evolved in environments very different from Earth's biological systems. Specifically, it proposes the idea of alien species emerging from non-biological processes in atmospheres that do not rely on the same types of biological energy resources that sustain life on this carbon-based habitable Earth.

►This part of the series briefly discusses the relationship between religion and energy levels, focusing on traditional practices and their interaction with modern brain systems. It also explores how the concept of an "Energy -God" fits into this framework. The series characterizes the "Energy-God" to help clarify the connections between faith and energy for individuals and living societies.

►All the books in the series include clear and reasonable drawings that demonstrate how energy functions in more than just specific directions. These drawings help explain that energy operates in several dimensions, not just the ones we're most familiar with, adding extra layers of complexity to how energy behaves in the actual Universe, especially in the matured 2nd-energy-belt.

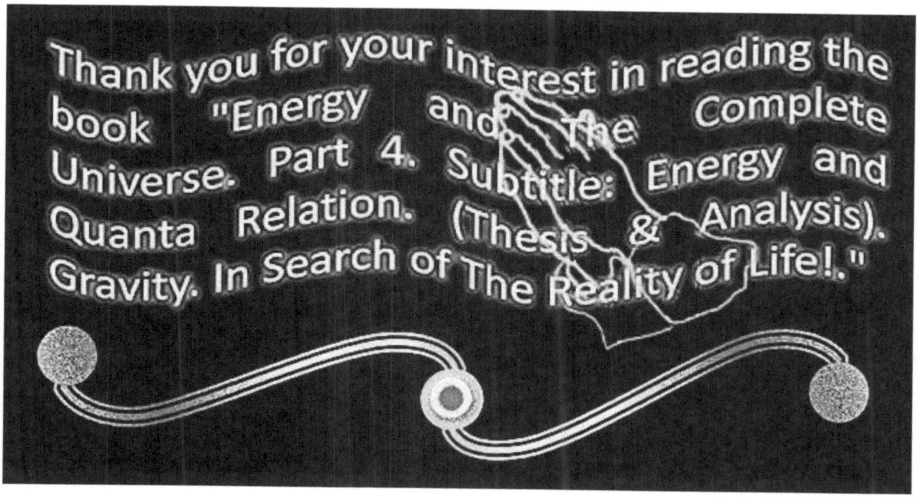
Thank you for your interest in reading the book "Energy and The Complete Universe. Part 4. Subtitle: Energy and Quanta Relation. (Thesis & Analysis). Gravity. In Search of The Reality of Life!."

9 789334 136081

Printed at Repro India Ltd.